THE GREAT MATHEMATICIANS

Professor Turnbull has written a biographical history
of mathematics from Thales and Pythagoras in the
sixth century B.C. to the great mathematical pioneers
of our own century.

To many people there is something forbidding
about the subject of mathematics; but the author has
succeeded in conveying to the reader the fascinating
interplay of numbers, problems and ideas that make
up the science and in describing the lives of the great
men who have devoted their lives to the first, and
most exacting of man's skills.

UNIVERSITY PAPERBACKS

UP 50

The Great Mathematicians

by

H. W. TURNBULL

UNIVERSITY PAPERBACKS

METHUEN : LONDON

First published 1929 by Methuen & Co
Fourth edition 1951
First published in this series 1962
Printed in Great Britain by
John Dickens and Co Ltd, Northampton
Catalogue No. 2/6800/27

The publishers are grateful to
the New York University Press
for permission to use the new
Introduction by James R. Newman

University Paperbacks are published by
METHUEN & CO LTD
36 Essex Street Strand WC2

Contents

v

Contents

Introduction

BY JAMES R. NEWMAN

H. W. Turnbull's excellent little book, a biographical history of mathematics, is the best work of its kind. It succeeds remarkably well in achieving the author's aim—to show through the representatives of their day in this venerable science "how a mathematician thinks, how his imagination, as well as his reason, leads him to new aspects of the truth."

Mathematics has been the subject of many histories. The classic works include Sir Thomas Heath's admirable two-volume *A History of Greek Mathematics,* his shorter *Manual,* his monographs on Apollonius, Archimedes, Aristarchus; Moritz Cantor's elaborate treatise, *Vorlesungen Über Geschichte Der Mathematik* (two volumes); Hieronymus Georg Zeuthen's *Geschichte Der Mathematik in Altertum und Mittelalter;* Florian Cajori's error-ridden, excessively condensed, dated, but still useful and widely used *A History of Mathematics;* David Eugene Smith's richly informative, well-illustrated two-volume *History of Mathematics,* a fine introduction to the elementary branches; Johannes Tropfke's seven-volume (when last heard from) *Geschichte der Elementarmathematik,* an enormously well-stocked survey; Gina Loria's *Storia Delle Matematiche,* a good three-volume account from antiquity to modern mathematics. Among the less formidable works are W. W. R. Ball's *A Short Account of the History of Mathematics,* and his *Primer,* both very popular, now old-fashioned but still enjoyable; J. W. N. Sullivan's *The History of Mathematics in Europe,* a readable outline, which carries the story only as far as the end of the eighteenth century; Dirk Struik's *A Concise History of Mathematics,* a book of solid merit but somewhat dull; J. E. Hofmann's three excellent German monographs, covering the subject from ancient times to the French Revolution; and J. F. Scott's *A History of Mathematics,* an interesting and able essay for teachers and students.

On the biographical side the late Eric Temple Bell's *Men of Mathematics* is of course unique, but at least two other less individualistic and colorful books may be mentioned: Alexander Macfarlane's *Lectures on Ten British Mathematicians,* and Ganesh Prasad's *Some Great Mathematicians of the Nineteenth Century.*

I have mentioned these books not so much for purposes of comparison as to provide a brief bibliographical run-down for those whose appetite may have been whetted by Turnbull. In my view the honors remain his for a brief, agreeable, instructive survey of men and ideas, accessible to the plain reader with enough curiosity and persistence not to be dismayed by high thoughts.

Herbert Westren Turnbull was born in Tettenhall, Wolverhampton, 1885. He was educated at Cambridge University, thereafter lectured on mathematics at Cambridge, Liverpool, and Hong Kong. On returning from the Far East he taught for a time at a public school and then became Regius Professor of Mathematics at United College, St. Andrews University. This post he held from 1921 to 1950 and was subsequently Professor Emeritus. He was elected a Fellow of the Royal Society in 1932. Turnbull was distinguished for his researches in algebra (determinants, matrices, theory of equations) and, as demonstrated not only in this volume but in other writings, was a gifted simplifier of mathematical ideas. The last years of his life he was the editor of the Royal Society's magnificent *The Correspondence of Isaac Newton,* of which three volumes have appeared, one of the most exciting and beautifully executed pieces of scholarship of recent years. Professor Turnbull died on May 4, 1961.

Preface

The usefulness of mathematics in furthering the sciences is commonly acknowledged: but outside the ranks of the experts there is little inquiry into its nature and purpose as a deliberate human activity. Doubtless this is due to the inevitable drawback that mathematical study is saturated with technicalities from beginning to end. Fully conscious of the difficulties in the undertaking, I have written this little book in the hope that it will help to reveal something of the spirit of mathematics, without unduly burdening the reader with its intricate symbolism. The story is told of several great mathematicians who are representatives of their day in this venerable science. I have tried to show how a mathematician thinks, how his imagination, as well as his reason, leads him to new aspects of the truth. Occasionally it has been necessary to draw a figure or quote a formula—and in such cases the reader who dislikes them may skip, and gather up the thread undismayed a little further on. Yet I hope that he will not too readily turn aside in despair, but will, with the help of the accompanying comment, find something to admire in these elegant tools of the craft.

Naturally in a work of this size the historical account is incomplete: a few references have accordingly been added for further reading. It is pleasant to record my deep obligation to the writers of these and other larger works, and particularly to my college tutor, the late Mr. W. W. Rouse Ball, who first woke my interest in the subject. My sincere thanks are also due to several former and present colleagues in St. Andrews who have made a considerable and illuminating study of mathematics among the Ancients: and to kind friends who have offered many valuable suggestions and criticisms.

In preparing the Second Edition I have had the benefit of suggestions which friends from time to time have submitted. I am grateful for this means of removing minor blemishes, and for making a few additions. In particular, a date list has been added.

PREFACE TO THIRD EDITION

A few additions have been made to the earlier chapters and to Chapter VI, which incorporate results of recent discoveries among mathematical inscriptions and manuscripts, particularly those which enlarge our knowledge of the mathematics of Ancient Babylonia and Egypt. I gratefully acknowledge the help derived from reading the Manual of Greek Mathematics *(1931) by Sir Thomas Heath. It provides a short but masterly account of these developments, for which the scientific world is greatly indebted.*

H. W. T.

December, 1940.

PREFACE TO FOURTH EDITION

At the turn of the half-century it is appropriate to add a postscript to Chapter XI, which brought the story of mathematical development as far as the opening years of the century. What has happened since has followed very directly from the wonderful advances that opened up through the algebraical discoveries of Hamilton, the analytical theories of Weierstrass and the geometrical innova-

tions of Von Staudt, and of their many great contemporaries. One very noteworthy development has been the rise of American mathematics to a place in the front rank, and this has come about with remarkable rapidity and principally through the study of abstract algebra such as was inspired by Peirce, a great American disciple of the Hamiltonian school. Representative of this advance in algebra is Wedderburn who built upon the foundations laid, not only by Peirce, but also by Frobenius in Germany and Cartan in France. Through abandoning the commutative law of multiplication by inventing quaternions, Hamilton had opened the door for the investigation of systems of algebra distinct from the ordinary familiar system. Algebra became algebras just as, through the discovery of non-Euclidean systems, geometry became geometries. This plurality, which had been unsuspected for so long, naturally led to the study of the classification of algebras. It was here that Wedderburn, following a hint dropped by Cartan, attained great success. The matter led to deeper and wider understanding of abstract theory, while at the same time it provided a welcome and fertile medium for the further developments in quantum mechanics. Simultaneously with this abstract approach to algebra a powerful advance was made in the technique of algebraical manipulation through the discoveries of Frobenius, Schur and A. Young in the theory of groups and of their representations and applications.

Similar trends may be seen in arithmetic and analysis where the same plurality is in evidence. Typical of this are the theory of valuation and the recognition of Banach spaces. The axiom of Archimedes (p. 31) is here in jeopardy: which is hardly surprising once the concept of regular equal steps upon a straight line had been broadened by the newer forms of geometry. Arithmetic and analysis were, so to speak, projected and made more abstract. It is remarkable that, with these trends towards generalization in each of the four great branches of pure mathematics, the branches lose something of their distinctive qualities and grow more alike. Whitehead's description of geometry as the science of cross-classification remains profoundly true. The applications of mathematics continue to extend, particularly in logic and in statistics.

H. W. T.

May, 1951.

Date List

? 18th Century B.C.			Ahmes (? 1800–).
6th	"	"	Thales (640–550), Pythagoras (569–500).
5th	"	"	Anaxagoras (? 500–428), Zeno (495–435), Hippocrates (470–), Democritus (? 470–).
4th	"	"	Archytas (? 400), Plato (429–348), Eudoxus (408–355), Menaechmus (? 375–325).
3rd	"	"	Euclid (? 330–275), Archimedes (287–212), Apollonius (? 262–200).
2nd	"	"	Hipparchus (? 160–).
1st	"	A.D.	Menelaus (? 100).
2nd	"	"	Ptolemy (? 100–168).
3rd	"	"	Hero (? 250), Pappus (? 300), Diophantus (– 320 ?).
6th	"	"	Arya-Bhata (? 530).
7th	"	"	Brahmagupta (? 640).
12th	"	"	Leonardo of Pisa (1175–1230).
16th	"	"	Scipio Ferro (1465–1526), Tartaglia (1500–1557), Cardan (1501–1576), Copernicus (1473–1543), Vieta (1540–1603), Napier (1550–1617), Galileo 1564–1642), Kepler (1571–1630), Cavalieri (1598–1647).
17th	"	"	Desargues (1593–1662), Descartes (1596–1650), Fermat (1601–1665), Pascal (1623–1662), Wallis (1616–1703), Barrow (1630–1677), Gregory (1638–1675), Newton (1642–1727), Leibniz (1646–1716), Jacob Bernoulli (1654–1705), John Bernoulli (1667–1748).
18th	"	"	Euler (1707–1783), Demoivre (1667–1754), Taylor (1685–1741), Maclaurin (1698–1746), D'Alembert (1717–1783), Lagrange (1736–1813), Laplace (1749–1827), Cauchy (1759–1857).
19th	"	"	Gauss (1777–1855), Von Staudt (1798–1867), Abel (1802–1829), Hamilton (1805–1865), Galois (1811–1832), Riemann (1826–1866), Sylvester (1814–1897), Cayley (1821–1895), Weierstrass (1815–1897), and many others.
20th	"	"	Ramanujan (1887–1920), and many living mathematicians.

CHAPTER I

Early Beginnings:
Thales, Pythagoras
and the Pythagoreans

To-day with all our accumulated skill in exact measurements, it is a noteworthy event when lines driven through a mountain meet and make a tunnel. How much more wonderful is it that lines, starting at the corners of a perfect square, should be raised at a certain angle and successfully brought to a point, hundreds of feet aloft! For this, and more, is what is meant by the building of a pyramid: and all this was done by the Egyptians in the remote past, far earlier than the time of Abraham.

Unfortunately we have no actual record to tell us who first discovered enough mathematics to make the building possible. For it is evident that such gigantic erections needed very accurate plans and models. But many general statements of the rise of mathematics in Egypt are to be found in the writings of Herodotus and other Greek travellers. Of a certain king Sesostris, Herodotus says:

'This king divided the land among all Egyptians so as to give each one a quadrangle of equal size and to draw from each his revenues, by imposing a tax to be levied yearly. But everyone from whose part the river tore anything away, had to go to him to notify what had happened; he then sent overseers who had to measure out how much the land had become smaller, in order that the owner might pay on what was left, in proportion to the entire tax imposed. In this way, it appears to me, geometry originated, which passed thence to Hellas.'

1

Then in the *Phaedrus* Plato remarks:

'At the Egyptian city of Naucratis there was a famous old god whose name was Theuth; the bird which is called the Ibis was sacred to him, and he was the inventor of many arts, such as arithmetic and calculation and geometry and astronomy and draughts and dice, but his great discovery was the use of letters.'

According to Aristotle, mathematics originated because the priestly class in Egypt had the leisure needful for its study; over two thousand years later exact corroboration of this remark was forthcoming, through the discovery of a papyrus, now treasured in the Rhind collection at the British Museum. This curious document, which was written by the priest AHMES, who lived before 1700 B.C., is called 'directions for knowing all dark things'; and the work proves to be a collection of problems in geometry and arithmetic. It is much concerned with the reduction of fractions such as $2/(2n + 1)$ to a sum of fractions each of whose numerators is unity. Even with our improved notation it is a complicated matter to work through such remarkable examples as:

$$\tfrac{2}{29} = \tfrac{1}{24} + \tfrac{1}{58} + \tfrac{1}{174} + \tfrac{1}{232}.$$

There is considerable evidence that the Egyptians made astonishing progress in the science of exact measurements. They had their land surveyors, who were known as *rope stretchers,* because they used ropes, with knots or marks at equal intervals, to measure their plots of land. By this simple means they were able to construct right angles; for they knew that three ropes, of lengths three, four and five units respectively, could be formed into a right-angled triangle. This useful fact was not confined to Egypt: it was certainly known in China and elsewhere. But the Egyptian skill in practical geometry went far beyond the construction of right angles:

2

for it included, besides the angles of a square, the angles of other regular figures such as the pentagon, the hexagon and the heptagon.

If we take a pair of compasses, it is very easy to draw a circle and then to cut the circumference into *six* equal parts. The six points of division form a regular hexagon, the figure so well known as the section of the honey cell. It is a much more difficult problem to cut the circumference into *five* equal parts, and a very much more difficult problem to cut it into *seven* equal parts. Yet those who have carefully examined the design of the ancient temples and pyramids of Egypt tell us that these particular figures and angles are there to be seen. Now there are two distinct methods of dealing with geometrical problems—the practical and the theoretical. The Egyptians were champions of the practical, and the Greeks of the theoretical method. For example, as Röber has pointed out, the Egyptians employed a practical rule to determine the angle of a regular heptagon. And although it fell short of theoretical precision, the rule was accurate enough to conceal the error, unless the figure were to be drawn on a grand scale. It would barely be apparent even on a circle of radius 50 feet.

Unquestionably the Egyptians were masters of practical geometry; but whether they knew the theory, the underlying reason for their results, is another matter. Did they know that their right-angled triangle, with sides of lengths three, four and five units, contained an *exact* right angle? Probably they did, and possibly they knew far more. For, as Professor D'Arcy Thompson has suggested, the very *shape* of the Great Pyramid indicates a considerable familiarity with that of the regular pentagon. A certain obscure passage in Herodotus can, by the slightest literal emendation, be made to yield excellent sense. It would imply that the area of each triangular face of the Pyramid is equal to the square of the vertical height; and

3

this accords well with the actual facts. If this is so, the ratios of height, slope and base can be expressed in terms of the 'golden section,' or of the radius of a circle to the side of the inscribed decagon. In short, there was already a wealth of geometrical and arithmetical results treasured by the priests of Egypt, before the early Greek travellers became acquainted with mathematics. But it was only after the keen imaginative eye of the Greek fell upon these Egyptian figures that they yielded up their wonderful secrets and disclosed their inner nature.

Among these early travellers was THALES, a rich merchant of Miletus, who lived from about 640 to 550 b.c. As a man of affairs he was highly successful: his duties as merchant took him to many countries, and his native wit enabled him to learn from the novelties which he saw. To his admiring fellow-countrymen of later generations he was known as one of the Seven Sages of Greece, many legends and anecdotes clustering round his name. It is said that Thales was once in charge of some mules, which were burdened with sacks of salt. Whilst crossing a river one of the animals slipped; and the salt consequently dissolving in the water, its load became instantly lighter. Naturally the sagacious beast deliberately submerged itself at the next ford, and continued this trick until Thales hit upon the happy expedient of filling the sack with sponges! This proved an effectual cure. On another occasion, foreseeing an unusually fine crop of olives, Thales took possession of every olive-press in the district; and having made this 'corner,' became master of the market and could dictate his own terms. But now, according to one account, as he had *proved* what could be done, his purpose was achieved. Instead of victimizing his buyers, he magnanimously sold the fruit at a price reasonable enough to have horrified the financier of to-day.

Like many another merchant since his time Thales early

retired from commerce, but unlike many another he spent his leisure in philosophy and mathematics. He seized on what he had learnt in his travels, particularly from his intercourse with the priests of Egypt; and he was the first to bring out something of the true significance of Egyptian scientific lore. He was both a great mathematician and a great astronomer. Indeed, much of his popular celebrity was due to his successful prediction of a solar eclipse in 585 B.C. Yet it is told of him that in contemplating the stars during an evening walk, he fell into a ditch; whereupon the old woman attending him exclaimed, 'How canst thou know what is doing in the heavens when thou seest not what is at thy feet?'

We live so far from these beginnings of a rational wonder at natural things, that we run the risk of missing the true import of results now so very familiar. There are the well-known propositions that a circle is bisected by any diameter, or that the angles at the base of an isosceles triangle are equal, or that the angle in a semicircle is a right angle, or that the sides about equal angles in similar triangles are proportional. These and other like propositions have been ascribed to Thales. Simple as they are, they mark an epoch. They elevate the endless details of Egyptian mensuration to general truths; and in like manner his astronomical results replace what was little more than the making of star catalogues by a genuine science.

It has been well remarked that in this geometry of Thales we also have the true source of algebra. For the theorem that the diameter bisects a circle is indeed a true equation; and in his experiment conducted, as Plutarch says, 'so simply, without any fuss or instrument' to determine the height of the Great Pyramid by comparing its shadow with that of a vertical stick, we have the notion of equal ratios, or proportion.

The very idea of abstracting all solidity and area from a material shape, such as a square or triangle, and pondering upon it as a pattern of lines, seems to be definitely due to Thales. He also appears to have been the first to suggest the importance of a geometrical *locus,* or curve traced out by a point moving according to a definite law. He is known as the father of Greek mathematics, astronomy and philosophy, for he combined a practical sagacity with genuine wisdom. It was no mean achievement, in his day, to break through the pagan habit of mind which concentrates on particular cults and places. Thales asserted the existence of the abstract and the more general: these, said he, were worthier of deep study than the intuitive or sensible. Here spoke the philosopher. On the other hand he gave to mankind such practical gifts as the correct number of days in the year, and a convenient means of finding by observation the distance of a ship at sea.

Thales summed up his speculations in the philosophical proposition 'All things are water.' And the fact that all things are not water is trivial compared with the importance of his outlook. He saw the field; he asked the right questions; and he initiated the search for underlying law beneath all that is ephemeral and transient.

Thales never forgot the debt that he owed to the priests of Egypt; and when he was an old man he strongly advised his pupil PYTHAGORAS to pay them a visit. Acting upon this advice, Pythagoras travelled and gained a wide experience, which stood him in good stead when at length he settled and gathered round him disciples of his own, and became even more famous than his master. Pythagoras is supposed to have been a native of Samos, belonging like Thales to the Ionian colony of Greeks planted on the western shores and islands of what we now call Asia Minor. He lived from about 584 B.C. to 495 B.C. In 529 B.C. he settled at Crotona, a town of the Dorian colony in South Italy, and there he began to lecture upon philosophy and mathematics. His lecture-room was

6

thronged with enthusiastic hearers of all ranks. Many of the upper classes attended, and even women broke a law which forbade them to attend public meetings, and flocked to hear him. Among the most attentive was Theano, the young and beautiful daughter of his host Milo, whom he married. She wrote a biography of her husband, but unfortunately it is lost.

So remarkable was the influence of this great master that the more attentive of his pupils gradually formed themselves into a society or brotherhood. They were known as the Order of the Pythagoreans, and they were soon exercising a great influence far across the Grecian world. This influence was not so much political as religious. Members of the Society shared everything in common, holding the same philosophical beliefs, engaging in the same pursuits, and binding themselves with an oath not to reveal the secrets and teaching of the school. When, for example, Hippasus perished in a shipwreck, was not his fate due to a broken promise? For he had divulged the secret of the sphere with its twelve pentagons!

A distinctive badge of the brotherhood was the beautiful star pentagram—a fit symbol of the mathematics which the school discovered. It was also the symbol of health. Indeed, the Pythagoreans were specially interested in the study of medicine. Gradually, as the Society spread, teachings once treasured orally were committed to writing. Thereby a copy of a treatise by Philolaus, we are told, ultimately came into the possession of Plato; probably a highly significant event in the history of mathematics.

In mathematics the Pythagoreans made very great progress, particularly in the theory of numbers and in the geometry of areas and solids. As it was the generous practice among members of the brotherhood to attribute all credit for each new discovery to Pythagoras himself, we cannot be quite certain about the authorship of each particular theorem. But at any

FIGURE 1

rate in the mathematics which are now to be described, his was the dominating influence.

In thinking of these early philosophers we must remember that open air and sunlight and starry nights formed their surroundings—not our grey mists and fettered sunshine. As Pythagoras was learning his mensuration from the priests of Egypt, he would constantly see the keen lines cast by the shadows of the pillars across the pavements. He trod chequered floors with their arrays of alternately coloured squares. His mind was stirred by interesting geometrical truths learnt from his master Thales, his interest in number would lead him to count the squares, and the sight of the long straight shadow falling obliquely across them would suggest sequences of special squares. It falls maybe across the centre of the first, the fourth, the seventh; the arithmetical progression is suggested. Then again the square is interesting for its *size*. A fragment of more diverse pattern would demonstrate a larger square enclosing one exactly half its size. A little imaginative thought would reveal, within the larger, a smaller square placed unsymmetrically, and so would lead to the great theorem which somehow or other was early reached by the brotherhood (and

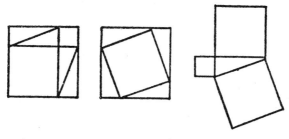

FIGURE 2

some say by Pythagoras himself), that the square on one side of a right-angled triangle is equal to the sum of the squares on the remaining sides. The above figures (Figure 2) actually suggest the proof, but it is quite possible that several different proofs were found, one being by the use of similar triangles. According to one story, when Pythagoras first discovered this fine result, in his exultation he sacrificed an ox!

Influenced no doubt by these same orderly patterns, he pictured numbers as having characteristic designs. There were the triangular numbers, one, three, six, ten, and so on, ten being the *holy tetractys,* another symbol highly revered by the

brotherhood. Also there were the square numbers, each of which could be derived from its predecessor by adding an L-

shaped border. Great importance was attached to this border: it was called a gnomon (γνώμων, carpenter's rule). Then it was recognized that each odd number, three, five, seven, etc., was a gnomon of a square number. For example, seven is the gnomon of the square of three to make the square of four. Pythagoras was also interested in the more abstract natural objects, and he is said to have discovered the wonderful harmonic progressions in the notes of the musical scale, by finding the relation between the length of a string and the pitch of its vibrating note. Thrilled by his discovery, he saw in numbers the element of all things. To him numbers were no mere attributes: three was not that which is common to three cats or three books or three Graces: but numbers were themselves the stuff out of which all objects we see or handle are made— the rational reality. Let us not judge the doctrine too harshly; it was a great advance on the cruder water philosophy of Thales.

So, in geometry, *one* came to be identified with the point; *two* with the line, *three* with the surface, and *four* with the solid. This is a noteworthy disposition that really is more fruitful than the usual allocation in which the line is said to have one, the surface two, and the solid three, dimensions.

More whimsical was the attachment of *seven* to the maiden goddess Athene 'because seven alone within the decade has neither factors nor product.' *Five* suggested marriage, the union of the first even with the first genuine odd number. *One* was further identified with reason; *two* with opinion—a wavering fellow is Two; he does not know his own mind: *four* with justice, steadfast and square. Very fanciful no doubt: but has not Ramanujan, one of the greatest arithmeticians of our own days, been thought to treat the positive integers as his personal friends? In spite of this exuberance the fact remains, as Aristotle sums it up: 'The Pythagoreans first applied themselves to mathematics, a science which they improved; and, penetrated with it, they fancied that the principles of mathe-

10

matics were the principles of all things.' And a younger contemporary, Eudemus, shrewdly remarked that 'they changed geometry into a liberal science; they diverted arithmetic from the service of commerce.'

To Pythagoras we owe the very word mathematics and its doubly two-fold branches:

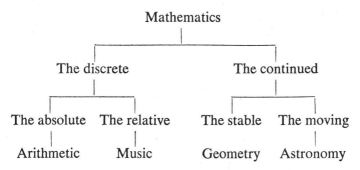

This classification is the origin of the famous Quadrivium of knowledge.

In geometry Pythagoras or his followers developed the theory of space-filling figures. The more obvious of these must have been very well known. If we think of each piece in such a figure as a unit, the question arises, can we fill a flat surface with repetitions of these units? It is very likely that this type of inquiry was what first led to the theorem that the three angles of a triangle are together equal to two right angles. The same train of thought also extends naturally to solid geometry, including the conception of regular solids. One of the diagrams (Figure 3) shows six equal triangles filling flat space round their central point. But five such equilateral triangles can likewise be fitted together, to form a blunted bell-tent-shaped figure, spreading from a central vertex: and now their bases form a regular pentagon. Such a figure is no longer flat; it makes a solid angle, the corner, in fact,

11

of a regular icosahedron. The process could be repeated by surrounding each vertex of the original triangles with five triangles. Exactly twenty triangles would be needed, no more and no less, and the result would be the beautiful figure of the

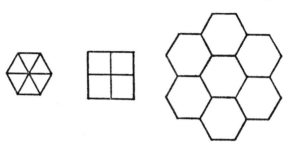

FIGURE 3

icosahedron of twenty triangles surrounding its twelve vertices in circuits of five.

It is remarkable that in solid geometry there are only five such regular figures, and that in the plane there is a very limited number of associations of regular space-filling figures. The three simplest regular solids, including the cube, were known to the Egyptians. But it was given to Pythagoras to discover the remaining two—the dodecahedron with its twelve pentagonal faces, and the icosahedron. Nowadays we so often become acquainted with these regular solids and plane figures only after a long excursion through the intricacies of mensuration and plane geometry that we fail to see their full simplicity and beauty.

Another kind of problem that interested Pythagoras was called the *method of application of areas*. His solution is noteworthy because it provided the geometrical equivalent of solving a quadratic equation in algebra. The main problem consisted in drawing, upon a given straight line, a figure that should be the size of one and the shape of another given figure. In the course of the solution one of three things was bound to happen. The base of the constructed figure would either fit, fall short of, or exceed the length of the given straight line. Pythagoras thought it proper to draw attention to these three possibilities; accordingly he introduced the terms *parabole, ellipsis* and *hyperbole*. Many years later his nomenclature was adopted by Apollonius, the great student of the conic section, because the same threefold characteristics presented themselves in the construction of the curve. And we who follow Apollonius still call the curve the parabola, the ellipse, or the hyperbola, as the case may be. The same threefold classification underlies the signs $=$, $<$, $>$ in arithmetic.

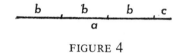

FIGURE 4

Many a time throughout the history of mathematics this classification has proved to be the key to further discoveries.

For example, it is closely connected with the theory of irrational numbers; and this brings us to the greatest achievement of Pythagoras, who is credited with discovering the (ἄλογον) irrational. In other words, he proved that it was not always possible to find a common measure for two given lengths a and b. The practice of measuring one line against another must have been very primitive. Here is a long line a, into which the shorter line b fits three times, with a still shorter piece c left over (Figure 4). To-day we express this by the equation $a = 3b + c$, or more generally by $a = nb + c$. If there is no such remainder c, the line b measures a; and a is called a multiple of b. If, however, there is a remainder c, further subdivision might perhaps account for each length a, b, c without remainder: experiment might show, for instance, that in tenths of inches, $a = 17$, $b = 5$, $c = 2$. At one time it was

13

thought that it was *always* possible to reduce lengths *a* and *b* to such multiples of a smaller length. It appeared to be simply a question of patient subdivision, and sooner or later the desired measure would be found. So the required subdivision, in the present example, is found by measuring *b* with *c*. For *c* fits twice into *b* with a remainder *d;* and *d* fits exactly twice into *c without* remainder. Consequently *d* measures *c,* and also measures *b* and also *a*. This is how the numbers 17, 5 and 2 come to be attached to *a, b,* and *c:* namely *a* contains *d* 17 times.

Incidentally this shows how naturally the arithmetical progression arises. For although the original subdivisions, and extremity, of the line *a* occur at distances 5, 10, 15, 17, measured from the left in quarter inches, they occur at distances 2, 7, 12, 17, from the right. These numbers form a typical arithmetical progression, with a rhythmical law of succession that alone would be incentive for a Pythagorean to study them further.

This reduction of the comparison of a line *a* with a line *b* to that of the number 17 with 5, or, speaking more technically, this reduction of the ratio *a* : *b* to 17 : 5 would have been agreeable to the Pythagorean. It exactly fitted in with his philosophy; for it helped to reduce space and geometry to pure number. Then came the awkward discovery, by Pythagoras himself, that the reduction was not always possible; that something in geometry eluded whole numbers. We do not know exactly how this discovery of the *irrational* took place, although two early examples can be cited. First when *a* is the diagonal and *b* is the side of a square, no common measure can be found; nor can it be found in a second example, when a line *a* is divided in *golden section* into parts *b* and *c*. By this is meant that the ratio of *a,* the whole line, to the part *b* is equal to the ratio of *b* to the other part *c*. Here *c* may be fitted once into *b* with remainder *d:* and then *d* may be fitted once into *c* with a remainder *e:* and so on. It is not hard to prove that such lengths *a, b, c, d, . . .* form a geometrical progression without end; and the desired common measure is never to be found.

If we prefer algebra to geometry we can verify this as follows. Since it is given that $a = b + c$ and also $a : b :: b : c$, it follows that $a(a - b) = b^2$. This is a quadratic equation for the ratio $a : b$, whose solution gives the result

$$a : b = \sqrt{5} + 1 : 2.$$

The presence of the surd $\sqrt{5}$ indicates the irrational.

The underlying reason why such a problem came to be studied is to be found in the star badge of the brotherhood (p. 8); for every line therein is divided in this golden section. The star has five lines, each cut into three parts, the lengths

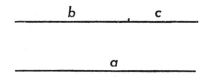

FIGURE 5

of which can be taken as *a, b, a.* As for the ratio of the diagonal to the side of a square, Aristotle suggests that the Pythagorean proof of its irrationality was substantially as follows:

If the ratio of diagonal to side is commensurable, let it be $p : q$, where p and q are whole numbers prime to one another. Then p and q denote the number of equal subdivisions in the diagonal and the side of a square respectively. But since the square on the diagonal is double that on the side, it follows that $p^2 = 2q^2$. Hence p^2 is an even number, and p itself must be even. Therefore p may be taken to be $2r$, p^2 to be $4r^2$, and

15

consequently q^2 to be $2r^2$. This requires q to be even; which is impossible because two numbers p, q, prime to each other cannot both be even. So the original supposition is untenable: no common measure can exist; and the ratio is therefore irrational.

This is an interesting early example of an indirect proof or *reductio ad absurdum;* and as such it is a very important step in the logic of mathematics.

We can now sum up the mathematical accomplishments of these early Greek philosophers. They advanced in geometry far enough to cover roughly our own familiar school course in the subject. They made substantial progress in the theoretical side of arithmetic and algebra. They had a geometrical equivalent for our method of solving quadratic equations; they studied various types of progressions, arithmetical, geometrical and harmonical. In Babylon, Pythagoras is said to have learnt the 'perfect proportion'

$$a : \frac{a + b}{2} :: \frac{2ab}{a + b} : b$$

which involves the arithmetical and harmonical means of two numbers. Indeed, to the Babylonians the Greeks owed many astronomical facts, and the sexagesimal method of counting by sixties in arithmetic. But they lacked our arithmetical notation and such useful abbreviations as are found in the theory of indices. From a present-day standpoint these results may be regarded as elementary: it is otherwise with their discovery of irrational numbers. That will ever rank as a piece of essentially advanced mathematics. As it upset many of the accepted geometrical proofs it came as a 'veritable logical scandal.' Much of the mathematical work in the succeeding era was coloured by the attempt to retrieve the position, and in the end this was triumphantly regained by Eudoxus.

Recent investigations of the Rhind Papyrus, the Moscow Papyrus of the Twelfth Egyptian Dynasty, and the Strassburg Cuneiform texts have greatly added to the prestige of Egyptian and Babylonian mathematics. While no general proof has yet been found among these sources, many remarkable *ad hoc* formulae have come to light, such as the Babylonian solution of complicated quadratic equations dating from 2000 B.C., which O. Neugebauer published in 1929, and an Egyptian approximation to the area of a sphere (equivalent to reckoning $\pi = 256/81$).

CHAPTER II

Eudoxus and
the Athenian School

A second stage in the history of mathematics occupied the fifth and fourth centuries B.C., and is associated with Athens. For after the wonderful victories at Marathon and Salamis early in the fifth century, when the Greeks defeated the Persians, Athens rose to a position of pre-eminence. The city became not only the political and commercial, but the intellectual centre of the Grecian world. Her philosophers congregated from East and West, many of whom were remarkable mathematicians and astronomers. Perhaps the greatest among these were Hippocrates, Plato, Eudoxus and Menaechmus; and contemporary with the three latter was Archytas the Pythagorean, who lived at Tarentum.

Thales and Pythagoras had laid the foundations of geometry and arithmetic. The Athenian school concentrated upon special aspects of the superstructure; and, whether by accident or design, found themselves embarking upon three great problems: (i) the *duplication of the cube,* or the attempt to find the edge of a cube whose volume is double that of a given cube; (ii) the *trisection of a given angle;* and (iii) the *squaring of a circle,* or, the attempt to find a square whose area is equal to that of a given circle. These problems would naturally present themselves in a systematic study of geometry; while, as years passed and no solutions were forthcoming they would attract increasing attention. Such is their inherent stubbornness that not until the nineteenth century were satisfactory answers to

18

these problems found. Their innocent enunciations are at once an invitation and a paradox. Early attempts to solve them led indirectly to results that at first sight seem to involve greater difficulties than the problems themselves. For example, in trying to square the circle Hippocrates discovered that two moon-shaped figures could be drawn whose areas were together equal to that of a right-angled triangle. This diagram (Figure 6) with its three semicircles described on the respective sides of the triangle illustrates his theorem. One might readily suppose that it would be easier to determine the area of a single circle than that of these lunes, or lunules, as they are called, bounded by pairs of circular arcs. Yet such is not the case.

In this by-product of the main problem Hippocrates gave the first example of a solution in *quadratures*. By this is meant the problem of constructing a rectilinear area equal to an area bounded by one or more curves. The sequel to attempts of this kind was the invention of the integral calculus by Archimedes, who lived in the next century. But his first success in the method was not concerned with the area of a circle, but with that of a parabola, a curve that had been dis-

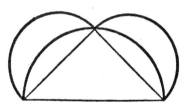

FIGURE 6

covered by Menaechmus in an attempt to duplicate the cube. This shows how very interdependent mathematics had now become with its interplay between branch and branch. All this activity led to the discovery of many other new curves, includ-

19

ing the ellipse, the hyperbola, the quadratrix, the conchoid (the shell), the cissoid (the ivy leaf), various spirals, and other curves classed as *loci on surfaces*.

The Greeks now found it useful to adopt a special classification for their problems, calling them *plane, solid* and *linear*. Problems were *plane* if their solution depended only on the use of straight lines and circles. These were of distinctly the Pythagorean type. They were *solid* if they depended upon conic sections: and they were *linear* if in addition they depended upon still more complicated curves. This early classification shows true mathematical insight, because later experience has revealed close algebraic and analytic parallels. For example, the *plane* problem corresponds in algebra to the problem soluble by quadratic equations. The Greeks quite naturally but vainly supposed that the three famous problems above were soluble by plane methods. It is here that they were wrong: for by solid or linear methods the problems were not necessarily insoluble.

One of the first philosophers to bring the new learning from Ionia to Athens was ANAXAGORAS (? 500–428 B.C.), who came from near Smyrna. He is said to have neglected his possessions, which were considerable, in order to devote himself to science, and in reply to the question, what was the object of being born, he remarked: 'The investigation of the sun, moon and heaven.' In Athens he shared the varying fortunes of his friend Pericles, the great statesman, and at one time was imprisoned for impiety. This we know from an ancient record which adds that 'while in prison he wrote (or drew) the squaring of the circle,' a brief but interesting allusion to the famous problem. Nor has the geometry of the circle suffered unduly from the captivity of its devotees. Centuries later another great chapter was opened, when the Russians flung Poncelet, an officer serving under Napoleon, into prison, where he discovered the circular points at infinity. Anaxagoras, however, was famous chiefly for his work in astronomy.

HIPPOCRATES* was his younger contemporary, who came from Chios to Athens about the middle of the fifth century. A lawsuit originally lured him to the city: for he had lost considerable property in an attack by Athenian pirates near Byzantium. Indeed, the tastes of Athenian citizens were varied: they were not all artists, sculptors, statesmen, dramatists, philosophers, or honest seamen, in spite of the wealth there and then assembled. After enduring their ridicule first at being cheated and then for hoping to recover his money, the simple-minded Hippocrates gave up the quest, and found his solace in mathematics and philosophy.

He made several notable advances. He was the first author who is known to have written an account of elementary mathematics; in particular he devoted his attention to properties of the circle. To-day his actual work survives among the theorems of Euclid, although his original book is lost. His chief result is the proof of the statement that circles are to one another in the ratio of the squares on their diameters. This is equivalent to the discovery of the formula πr^2 for the area of a circle in terms of its radius. It means that a certain number π exists, and is the same for all circles, although his method does not give the actual numerical value of π. It is thought that he reached his conclusions by looking upon a circle as the limiting form of a regular polygon, either inscribed or circumscribed. This was an early instance of the *method of exhaustions*—a particular use of approximation from below and above to a desired limit.

The introduction of the method of exhaustions was an important link in the chain of thought culminating in the work of Eudoxus and Archimedes. It brought the prospect of un-

* Not the great physician.

ravelling the mystery of irrational numbers, that had sorely puzzled the early Pythagoreans, one stage nearer. A second important but perhaps simpler work of Hippocrates was an example of the useful device of reducing one theorem to another. The Pythagoreans already had shown how to find the geometric mean between two magnitudes by a geometrical construction. They merely drew a square equal to a given rectangle. Hippocrates now showed that to duplicate a cube was tantamount to finding *two* such geometric means. Put into more familiar algebraic language, if

$$a : x = x : b, \text{ then } x^2 = ab,$$

and if

$$a : x = x : y = y : 2a$$

then $x^3 = 2a^3$. Consequently if a was the length of the edge of the given cube, x would be that of a cube twice its size. But the statement also shows that x is the first of two geometric means between a and $2a$.

We must, of course, bear in mind that the Greeks had no such convenient algebraic notation as the above. Although they went through the same reasoning and reached the same conclusions as we can, their statements were prolix, and afforded none of the help which we find in these concise symbols of algebra.

It is supposed that the study of the properties of two such means, x and y, between given lergths a and b, led to the discovery of the parabola and hyperbola. As we should say, nowadays, the above continued proportions indicate the equations $x^2 = ay$, and $xy = 2a^2$. These equations represent a parabola and a hyperbola: taken together they determine a point of intersection which is the key to the problem. This is an instance of a *solid* solution for the duplication of the cube. It represents the ripe experience of the Athenian school; for MENAECHMUS (? 375–325 B.C.), to whom it is credited, lived a century later than Hippocrates.

Where two lines, straight or curved, cross, is a point: where three surfaces meet is a point. The two walls and the ceiling meeting at the corner of a room give a convenient example. But two curved walls, meeting a curved ceiling would also make a corner, and in fact illustrate a truly ingenious method of dealing with this same problem of the cube. The author of this geometrical novelty was ARCHYTAS (? 400 B.C.), a contemporary of Menaechmus. This time the problem was reduced to finding the position of a certain point in space: and the point was located as the meeting-place of three surfaces. For one surface Archytas chose that generated by a circle revolving about a fixed tangent as axis. Such a surface can be thought of as a ring, although the hole through the ring is completely stopped up. His other surfaces were more commonplace, a cylinder and a cone. With this unusual choice of surfaces he succeeded in solving the problem. When we bear in mind how little was known, in his day, about solid geometry, this achievement must rank as a gem among mathematical antiquities. Archytas, too, was one of the first to write upon mechanics, and he is said to have been very skilful in making toys and models—a wooden dove which could fly, and a rattle which, as Aristotle says, 'was useful to give to children to occupy them from breaking things about the house (for the young are incapable of keeping still).'

Unlike the majority of mathematicians who lived in this Athenian era, Archytas lived at Tarentum in South Italy. He found time to take a considerable part in the public life of his city, and is known for his enlightened attitude in his treatment of slaves and in the education of children. He was a Pythagorean, and was also in touch with the philosophers of Athens, numbering Plato among his friends. He is said upon one occasion to have used his influence in high quarters to save the life of Plato.

Between Crotona and Tarentum upon the shore of the gulf of Southern Italy was the city of Elea: and with each of these places we may associate a great philosopher or mathematician. At Crotona Pythagoras had instituted his lecture-room; nearly two centuries later Archytas made his mechanical models at Tarentum. But about midway through the intervening period there lived at Elea the philosopher ZENO. This acutely original thinker played the part of philosophical critic to the mathematicians, and some of his objections to current ideas about motion and the infinitesimal were very subtle indeed. For example, he criticized the infinite geometrical progression by proposing the well-known puzzle of Achilles and the tortoise. How, asked Zeno, can the swift Achilles overtake the tortoise if he concedes a handicap? For if Achilles starts at A and the tortoise at B, then when Achilles reaches B the tortoise is at C, and when Achilles reaches C the tortoise is at D. As this description can go on and on, apparently Achilles never overtakes the tortoise. But actually he may do so; and this is a paradox. The point of the inquiry is not *when,* but *how* does Achilles overtake the tortoise.

Somewhat similar questions were asked by DEMOCRITUS, the great philosopher of Thrace, who was a contemporary of Archytas and Plato. Democritus has long been famous as the originator of the atomic theory, a speculation that was immediately developed by Epicurus, and later provided the great theme for the Latin poet Lucretius. It is, however, only quite recently that any mathematical work of Democritus has come to light. This happened in 1906, when Heiberg discovered a lost book of Archimedes entitled the *Method.* We learn from it that Archimedes regarded Democritus as the first

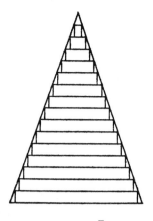

FIGURE 7

mathematician to state correctly the formula for the volume of a cone or a pyramid. Each of these volumes was one-third part of a circumscribing cylinder, or prism, standing on the same base. To reach his conclusions, Democritus thought of these solids as built up of innumerable parallel layers. There would be no difficulty in the case of the cylinder, for each layer would be equal. But for the cone or pyramid the sizes of layer upon layer would taper off to a point. The appended diagram (Figure 7), showing the elevation of a cone or pyramid, illustrates this tapering of the layers, although the picture that Democritus had in mind consisted of very much thinner layers. He was puzzled by their diminishing sizes. 'Are they equal or unequal?' he asked, 'for if they are unequal, they will make the cone irregular as having many indentations, like steps, and unevennesses; but, if they are equal, the sections will be equal, and the cone will appear to have the property of the cylinder and to be made up of equal, not unequal circles, which is very absurd.'

This quotation is striking; for it foreshadows the great constructive work of Archimedes, and, centuries later, that of Cavalieri and Newton. It exhibits the infinitesimal calculus in its infancy. The notion of stratification—that a solid could be thought of as layer upon layer—would occur quite naturally to Democritus, because he was a physicist; it would *not* so readily have occurred to Pythagoras or Plato with their more algebraic turn of mind which attracted them to the pattern or arrangement of things. But here the acute Greek thought is once more restless. No mere rough and ready approximation will satisfy Democritus: there is discrepancy between stratified pyramid and smoothly finished whole. The deep question of the theory of limits is at issue; but how far he foresaw any solution, we do not know.

This brings us to the great arithmetical work at Athens, associated with the names of PLATO (429–348 B.C.) and EUDOXUS (408–355 B.C.).

Among the philosophers of Athens only two were native to the place, Socrates and Plato, master and disciple, both of whom were well-read mathematicians. Plato was perhaps an original investigator; but whether this is so or not, he exerted an immense influence on the course that mathematics was to take, by founding and conducting in Athens his famous Academy. Over the entrance of his lecture room his students read the telling inscription, 'Let no one destitute of geometry enter my doors'; and it was his earnest wish to give his pupils the finest possible education. A man, said he, should acquire no mere bundle of knowledge, but be trained to see below the surface of things, seeking rather for the eternal reality and the Good behind it all. For this high endeavour the study of mathematics is essential; and numbers, in particular, must be studied, simply as numbers and not as embodied in anything. They impart a character to nature; for instance, the periods

26

of the heavenly bodies can only be characterized by invoking the use of irrationals.

Originally the Greek word ἀριθμοί, from which we derive 'arithmetic,' meant the natural numbers, although it was at first questioned whether unity was a number; for 'how can unity, the measure, be a number, the thing measured?' But by including irrationals as numbers Plato made a great advance: he was in fact dealing with what we nowadays call the positive real numbers. Zero and negative numbers were proposed at a far later date.

There is grandeur here in the importance which Plato ascribes to arithmetic for forming the mind: and this is matched by his views on geometry, 'the subject which has very ludicrously been called mensuration' (γεωμετρια = land measuring) but which is really an art, a more than human miracle in the eyes of those who can appreciate it. In his book, the *Timaeus*, where he dramatically expounds the views of his hero Timaeus, the Pythagorean, reference is made to the five regular solids and to their supposed significance in nature. The speaker tells how that the four elements earth, air, fire and water have characteristic shapes: the cube is appropriated to earth, the octahedron to air, the sharp pyramid or tetrahedron to fire, and the blunter icosahedron to water, while the Creator used the fifth, the dodecahedron, for the Universe itself. Is it sophistry, or else a brilliant foretaste of the molecular theory of our own day? According to Proclus, the late Greek commentator, 'Plato caused mathematics in general and geometry in particular to make a very great advance, by reason of his enthusiasm for them, which of course is obvious from the way he filled his books with mathematical illustrations, and everywhere tries to kindle admiration for these subjects, in those who make a pursuit of philosophy.' It is related that to the question, What does God do? Plato replied, 'God always geometrizes.'

Among his pupils was a young man of Cnidus, named Eudoxus, who came to Athens in great poverty, and, like many another poor student, had a struggle to maintain himself. To relieve his pocket he lodged down by the sea at the Piraeus, and every day used to trudge the dusty miles to Athens. But his genius for astronomy and mathematics attracted attention and finally brought him to a position of eminence. He travelled and studied in Egypt, Italy and Sicily, meeting Archytas, the geometer, and other men of renown. About 368 B.C., at the age of forty, Eudoxus returned to Athens in company with a considerable following of pupils, about the time when Aristotle, then a lad of seventeen, first crossed the seas to study at the Academy.

In astronomy the great work of Eudoxus was his theory of concentric spheres explaining the strange wanderings of the planets; an admirable surmise that went far to fit the observed facts. Like his successor Ptolemy, who lived many centuries later, and all other astronomers until Kepler, he found in circular motion a satisfactory basis for a complete planetary theory. This was great work; yet it was surpassed by his pure mathematics which touched the zenith of Greek brilliance. For Eudoxus placed the doctrine of irrationals upon a thoroughly sound basis, and so well was his task done that it still continues, fresh as ever, after the great arithmetical reconstructions of Dedekind and Weierstrass during the nineteenth century. An immediate effect of the work was to restore confidence in the geometrical methods of proportion and to complete the proofs of several important theorems. The *method of exhaustions* vaguely underlay the results of Democritus upon the volume of a cone and of Hippocrates on the area of a circle. Thanks to Eudoxus this method was fully explained.

An endeavour will now be made to indicate in a simple

way how this great object was achieved. This study of higher arithmetic at Athens was stimulated by the Pythagorean, Theodorus of Cyrene, who is said to have been Plato's teacher. For Theodorus discovered many irrationals, $\sqrt{3}$, $\sqrt{5}$, $\sqrt{6}$, $\sqrt{7}$, $\sqrt{8}$, $\sqrt{10}$, $\sqrt{11}$, $\sqrt{12}$, $\sqrt{13}$, $\sqrt{14}$, $\sqrt{15}$, $\sqrt{17}$, 'at which point,' says Plato, 'for some reason he stopped.' The omissions in the list are obvious: $\sqrt{2}$ had been discovered by Pythagoras through the ratio of diagonal to side of a square, while $\sqrt{4}$, $\sqrt{9}$, $\sqrt{16}$ are of course irrelevant. Now it is one thing to discover the *existence* of an irrational such as $\sqrt{2}$; it is quite another matter to find a way of *approach* to the number. It was this second problem that came prominently to the fore: it provided the arithmetical aspect of the *method of exhaustions* already applied to the circle: and it revealed a wonderful example of ancient arithmetic. We learn the details from a later commentator, Theon of Smyrna.

Unhampered by a decimal notation (which here is a positive hindrance, useful as it is in countless other examples), the Greeks set about their task in the following engaging fashion. To approximate to $\sqrt{2}$ they built a ladder of whole numbers. A brief scrutiny of the ladder shows how the rungs are devised: $1 + 1 = 2$, $1 + 2 = 3$, $2 + 3 = 5$, $2 + 5 = 7$, $5 + 7 = 12$, and so on. Each rung of the ladder consists of two numbers x and y, whose ratio approaches nearer and nearer to the ratio $1:\sqrt{2}$, the further down the ladder it is

1	1
2	3
5	7
12	17
29	41
etc.	

situated. Again, these numbers x and y, at each rung, satisfy the equation

$$y^2 - 2x^2 = \pm 1.$$

The positive and negative signs are taken at alternate rungs, starting with a negative. For example, at the third rung $7^2 - 2 \cdot 5^2 = -1$.

As these successive ratios are alternately less than and

greater than all that follow, they nip the elusive limiting ratio $1:\sqrt{2}$ between two extremes, like the ends of a closing pair of pincers. They approximate *from both sides* to the desired irrational: $\frac{5}{7}$ is a little too large, but $\frac{12}{17}$ is a little too small. Like pendulum swings of an exhausted clock they die down— but they never actually come to rest. Here again, is the Pythagorean notion of *hyperbole* and *ellipsis;* it was regarded as very significant, and was called by the Greeks the 'dyad' of the 'great and small.'

Such a ladder could be constructed for any irrational; and another very pretty instance, which has been pointed out by Professor D'Arcy Thompson, is closely connected with the problem of the Golden Section. Here the right member of

1	1
1	2
2	3
3	5
5	8
etc.	

each rung is the sum of the pair on the preceding rung, so that the ladder may be extended with the greatest ease. In this case the ratios approximate, again by the little more and the little less, to the limit $\sqrt{5}+1:2$. It is found that they provide the arithmetical approach to the golden section of a line AB, namely when C divides AB so that $CB:AC = AC:AB$.

FIGURE 8

In fact, AC is roughly $\frac{3}{5}$ of the length AB, but more nearly $\frac{5}{8}$ of AB; and so on. It is only fair to say that this simplest of all such ladders has not yet been found in the ancient literature, but owing to its intimate connexion with the pentagon, it is difficult to resist the conclusion that the later Pythagoreans were familiar with it. The series 1, 2, 3, 5, 8, . . . was known in mediaeval times to Leonardo of Pisa, surnamed Fibonacci, after whom it is nowadays named.

Let us now combine this ladder-arithmetic with the geometry of a divided line. For example, let a line AB be divided at random by C, into lengths *a* and *b*, where AC = *a*, CB = *b*. Then the question still remains, what is the *exact* arithmetical meaning of the ratio *a : b*, whether or not this is irrational? The wonderful answer to this question is what has made Eudoxus so famous. Before considering it, let us take as an illustration the strides of two walkers. A tall man A takes a regular stride of length *a*, and his short friend B takes a stride *b*. Now suppose that eight strides of A cover the same ground as thirteen of B: in this case the *single* strides of A and B are in the ratio 13 : 8. The repetition of strides, to make them cover a considerable distance, acts as a magnifying glass and helps in the measurement of the *single* strides *a* and *b*, one against the other. Here we have the point of view adopted by Eudoxus. Let us, says he in effect, multiply our magnitudes *a* and *b*, whose ratio is required, and see what happens.

Let us, he continues, be able to recognize if *a* and *b* are equal, and if not, which is greater. Then if *a* is the greater, let us secondly be able to find multiples 2*b*, 3*b*, . . . , *nb*, of the smaller magnitude *b;* and thirdly, let us always be able to find a multiple *nb* of *b* which exceeds *a*. (The tall man may have seven-league boots and the short man may be Tom Thumb. Sooner or later the dwarf will be able to overtake one stride of his friend!) Few will gainsay the propriety of these mild assumptions: yet their mathematical implications have proved to be very subtle. This third supposition of Eudoxus has been variously credited, but to-day it is known as the *Axiom of Archimedes*.

A definition of equal ratios can now be stated. Let *a, b, c, d* be four given magnitudes, then the ratio *a : b* is equal to that of *c : d*, if, *whatever* equimultiples *ma, mc* are chosen and whatever equimultiples *nb, nd* are chosen,

either ma > nb, mc > nd, (i)
or ma = nb, mc = nd, (ii)
or ma < nb, mc < nd. (iii)

On this strange threefold statement the whole theory of proportion for geometry and algebra was reared. It is impossible to develop the matter here in any convincing way, but the simplicity of the ingredients in this definition is remarkable enough to merit attention. It has the characteristic threefold pattern already noticed by Pythagoras. As far as ordinary commensurable ratios go, the statement (ii) would suffice; m and n are whole numbers and the ratios $a : b$, $c : d$ are each equal to the ratio $n : m$. But the essence of the new theory lies in (i) and (iii), because (ii) *never* holds for incommensurables—the geometrical equivalent of irrationals in arithmetic. But it is extraordinary that out of these inequalities *equal* ratios emerge.

Lastly it was a stroke of genius when Eudoxus put on record the above Axiom of Archimedes. To continue our illustration, marking time is not striding, and Eudoxus excluded marking time. However small the stride b might be, it had a genuine length. Eudoxus simply ruled out the case of a ratio $a : b$ when either a or b was zero. Thereby he avoided a trap that Zeno had already set, and into which many a later victim was to fall. So the axiom was a notice-board to warn the unwary. It also had another use; it automatically required a and b to be magnitudes of the same kind. For if a denoted length and b weight, no number of ounces could be said to exceed the length of a foot.

The logical triumphs of this great period in Grecian mathematics overshadow important but less spectacular advances which were made in numerical notation and in music. From the earliest times the significance of the numbers five and ten for counting had been recognised in Babylonia, China and Egypt: and in Homer πεμπάζειν 'to five' means to count. Eventually the Greeks systematised their written notation by using the letters of the alphabet to denote definite numbers ($\alpha = 1$, $\beta = 2$, $\gamma = 3$ and so on). A Halicarnassus inscription (circa 450 B.C.) provides perhaps the earliest attested use of this alphabetical numeration.

In music Archytas gave the numerical ratios for the intervals of the tetrachord on three scales, the enharmonic, the chromatic and the diatonic. He held that sound was due to impact, and that higher tones correspond to quicker, and lower tones to slower, motion communicated to the air.

CHAPTER III

Alexandria: Euclid, Archimedes and Apollonius

Towards the end of the fourth century B.C., the scene of mathematical learning shifted from Europe to Africa. By an extraordinary sequence of brilliant victories the young soldier-prince, Alexander of Macedonia, conquered the Grecian world, and conceived the idea of forming a great empire. But he died at the age of thirty-three (323 B.C.), only two years after founding the city of Alexandria. He had planned this stronghold near the mouth of the Nile on a magnificent scale, and the sequel largely fulfilled his hopes. Geographically it was a convenient meeting-place for Greek and Jew and Arab. There, what was finest in Greek philosophy was treasured in great libraries: the mathematics of the ancients was perfected: the intellectual genius of the Greek came into living touch with the moral and spiritual genius of the Jew: the Septuagint translation of Old Testament Scriptures was produced: and in due time it was there that the great philosophers of the early Christian Church taught and prospered. In spite of ups and downs the city endured for about six hundred years, but suffered grievous losses in the wilder times that followed. The end came in A.D. 642, when a great flood of Arab invasion surged westward, and Alexandria fell into the hands of the Calif Omar.

A great library, reputed to hold 700,000 volumes, was lost or destroyed in this series of disasters. But happily a remnant of its untold wealth filtered through to later days when the

Arabs, who followed the original warriors, came to appreciate the spoils upon which they had fallen. Ptolemy, the successor of Alexander in his African dominions, had founded this library about 300 B.C. He had in effect established a University; and among the earliest of the teachers was EUCLID. We know little of his life and character, but he most probably passed his years of tuition at Athens before accepting the invitation of Ptolemy to settle in Alexandria. For twenty or thirty years he taught, writing his well-known *Elements* and many other works of importance. This teaching bore notable fruit in the achievements of Archimedes and Apollonius, two of the greatest members of the University.

The picture has been handed down of a genial man of learning, modest and scrupulously fair, always ready to acknowledge the original work of others, and conspicuously kind and patient. Some one who had begun to read geometry with Euclid, on learning the first theorem asked, 'What shall I get by learning these things?' Euclid called his slave and said, 'Give him threepence, since he must make gain out of what he learns.' Apparently Euclid made much the same impression as he does to-day. The schoolboy, for whom the base angles of an isosceles triangle 'are forced to be equal, without any nasty proof,' is but re-echoing the ancient critic who remarked that two sides of a triangle were greater than the third, as was evident to an ass. And no doubt they told Euclid so.

In the *Elements* Euclid set about writing an exhaustive account of mathematics—a colossal task even in his day. The Work consisted of thirteen books, and the subjects of several books are extremely well known. Books I, II, IV, VI on lines, areas and simple regular plane figures are mostly Pythagorean, while Book III on circles expounds Hippocrates. The lesser known Book V elaborates the work of Eudoxus on proportion, which was needed to justify the properties of similar figures discussed in Book VI. Books VII, VIII and IX are arithmetical, giving an interesting account of the theory of numbers; and again much here is probably Pythagorean. Prime and composite numbers are introduced—a relatively late distinction; so are the earlier G.C.M. and L.C.M. of numbers, the theory of geometrical progressions, and in effect the theorem $a^{m+n} = a^m a^n$, together with a method of summing the progression by a beautiful use of equal ratios. Incidentally Euclid utilized this method to give his *perfect* numbers, such as 6, 28, 496, each of which is equal to the sum of its factors. The collection of perfect numbers still interests the curious; they are far harder to find than the rarest postage stamps. The ninth specimen alone has thirty-seven digits, while a still larger one is $2^{126}(2^{127} - 1)$.

Book X of Euclid places the writer in the forefront among analysts. It is largely concerned with the doctrine of irrational numbers, particularly of the type $\sqrt{(\sqrt{a} \pm \sqrt{b})}$, where a and b are positive integers. Here Euclid elaborates the arithmetical side of the work of Eudoxus, having already settled the geometrical aspect in Books V and VI, and here we duly find the method of exhaustions carefully handled. After Book XI on elementary solid geometry comes the great Book XII, which illustrates the same method of exhaustions by formally proving Hippocrates' theorem for πr^2, the area of a circle. Finally in Book XIII we have the climax to which all this stately procession has been leading. The Greeks were never in a hurry; and it is soothing, in these days of bustle, to contemplate the working of their minds. This very fine book gives and proves the constructions for the five regular solids of Pythagoras, extolled by Plato; and it ends with the dodecahedron, the symbol of the Universe itself.

By this great work Euclid has won the admiration and

helped to form the minds of all his successors. To be sure a few logical blemishes occur in his pages, the gleanings of centuries of incessant criticism; but the wonder is that so much has survived unchanged. In point of form he left nothing to be desired, for he first laid down his careful definitions, then his common assumptions or axioms, and then his postulates, before proceeding with the orderly arrangement of their consequences. There were, however, certain gaps and tautologies among these preliminaries of his work: they occur in the geometrical, not in the Eudoxian parts of his books; and it has been one of the objects of latter-day criticism to supply what Euclid left unsaid.

But on one point Euclid was triumphant; in his dealing with parallel lines. For he made no attempt to hide, by a plausible axiom, his inability to prove a certain property of coplanar lines. Most of his other assumptions, or necessary bases of his arguments, were such as would reasonably command general assent. But in the case of parallel lines he started with the following elaborate supposition, called the *Parallel Postulate:*

If a straight line meet two straight lines, so as to make the two interior angles on the same side of it taken together less than two right angles, these straight lines, being continually produced, shall at length meet on that side on which are the angles which are less than two right angles.

By leaving this unproved, and by actually proving its converse, Euclid laid himself open to ridicule and attack. Surely, said the critics, this is no proper assumption; it must be capable of proof. Hundreds of attempts were vainly made to remove this postulate by proving its equivalent; but each so-called proof carried a lurking fallacy. The vindication of Euclid came with the discovery in the nineteenth century of non-Euclidean geometry, when fundamental reasons were found for some such postulate. There is dignity in the way that Euclid left this curious rugged excrescence, like a natural outcrop of rock in the plot of ground that otherwise had been so beautifully smoothed.

Many of his writings have come down to us, dealing with astronomy, music and optics, besides numerous other ways of treating geometry in his *Data* and *Division of Figures*. But his *Book of Fallacies* with its intriguing title, and the *Porisms* are lost; and we only learn of them indirectly through Pappus, another great commentator. It is one of the historical puzzles of mathematics to discover what porisms were, and many geometers, notably Simson in Scotland and Chasles in France, have tried to do so. Very likely they were properties relating to the organic description of figures—a type of geometry that appealed to Newton, Maclaurin, and to workers in projective geometry of our own days. Geometry at Alexandria was in fact a wide subject, and it has even been thought by some that the *porisms* consisted of an analytical method, foreshadowing the co-ordinate geometry of Descartes.

Euclid was followed by ARCHIMEDES of Samos, and APOLLONIUS of Perga. After the incomparable discoveries of Eudoxus, so well consolidated by Euclid, it was now the time for great constructive work to be launched; and here were the giants to do it. Archimedes, one of the greatest of all mathematicians, was the practical man of common sense, the Newton of his day, who brought imaginative skill and insight to bear upon metrical geometry and mechanics, and even invented the integral calculus. Apollonius, one of the greatest of geometers, endowed with an eye to see form and design, followed the lead of Menaechmus, and perfected the geometry of conic sections. They sowed in rich handfuls the seeds of pure mathematics, and in due time the harvest was ingathered by Kepler and Newton.

38

Little is known of the outward facts in the life of Archimedes. His father was Phidias the astronomer, and he was possibly related to Hiero II, King of Syracuse, who certainly was his friend. As a youth Archimedes spent some time in Egypt, presumably at Alexandria with the immediate successors of Euclid—perhaps studying with Euclid himself. Then on returning home he settled in Syracuse, where he made his great reputation. In 212 B.C., at the age of seventy-five, he lost his life in the tumult that followed the capture of Syracuse by the Romans. Rome and Carthage were then at grips in the deadly Punic wars, and Sicily with its capital Syracuse lay as a 'No man's land' between them. During the siege of Syracuse by the Romans, Archimedes directed his skill towards the discomfiture of the enemy, so that they learnt to fear the machines and contrivances of this intrepid old Greek. The story is vividly told by Plutarch, how at last Marcellus, the Roman leader, cried out to his men, 'Shall we not make an end of fighting against this geometrical Briareus who uses our ships like cups to ladle water from the sea, drives off our sambuca ignominiously with cudgel-blows, and by the multitude of missiles that he hurls at us all at once, outdoes the hundred-handed giants of mythology!' But all to no purpose, for if the soldiers did but see a piece of rope or wood projecting above the wall, they would cry, 'There it is,' declaring that Archimedes was setting some engine in motion against them, and would turn their backs and run away. Of course the geometrical Briareus attached no importance to these toys; they were but the diversions of geometry at play. Ignoble and sordid, unworthy of written record, was the business of mechanics and every sort of art which was directed to use and profit. Such was the outlook of Archimedes.

He held these views to the end; for even after the fall of the city he was still pondering over mathematics. He had drawn a diagram in the sand on the ground and stood lost in thought, when a soldier struck him down. As Whitehead has remarked:

'The death of Archimedes at the hands of a Roman soldier is symbolical of a world change of the first magnitude. The Romans were a great race, but they were cursed by the sterility which waits upon practicality. They were not dreamers enough to arrive at new points of view, which could give more fundamental control over the forces of nature. No Roman lost his life because he was absorbed in the contemplation of a mathematical diagram.'

Many, but not all, of the wonderful writings of Archimedes still survive. They cover a remarkable mathematical range, and bear the incisive marks of genius. It has already been said that he invented the integral calculus. By this is meant that he gave strict proofs for finding the areas, volumes and centres of gravity of curves and surfaces, circles, spheres, conics, and spirals. By his method of finding a tangent to a spiral he even embarked on what is nowadays called differential geometry. In this work he had to invoke algebraic and trigonometric formulae; here, for example, are typical results:

$$1^2 + 2^2 + 3^2 + \ldots + n^2 = \tfrac{1}{6}n(n+1)(2n+1),$$

$$\sin \frac{\pi}{2n} + \sin \frac{2\pi}{2n} + \ldots + \sin (2n-1)\frac{\pi}{2n} = \cot \frac{\pi}{4n}.$$

This last is the concise present-day statement of a geometrical theorem, arising in his investigation of the value of π, which he gave approximately in various ways, such as

$$3\tfrac{1}{7} > \pi > 3\tfrac{10}{71}.$$

Elsewhere he casually states approximations to $\sqrt{3}$ in the form

$$\tfrac{265}{153} < \sqrt{3} < \tfrac{1351}{780},$$

40

which is an example of the ladder-arithmetic of the Pythagoreans (p. 29). As these two fractions are respectively equal to

$$\tfrac{1}{3}\left(5 + \cfrac{1}{5 + \tfrac{1}{10}}\right) \text{ and } \tfrac{1}{3}\left(5 + \cfrac{1}{5 + \cfrac{1}{10 + \tfrac{1}{5}}}\right),$$

it is natural to suppose that Archimedes was familiar with continued fractions, or else with some virtually equivalent device; especially as $\sqrt{3}$ itself is given by further continuation of this last fraction, with denominators 10, 5, 10, 5 in endless succession. The same type of arithmetic occurs elsewhere in his writings, as well as in those of his contemporary, Aristarchus of Samos, a great astronomer who surmised that the earth travels round the sun.

Allusion has already been made to the recent discovery of the *Method* of Archimedes, a book that throws light on the mathematical powers of Democritus. It also reveals Archimedes in a confidential mood, for in it he lifts the veil and tells us how some of his results were reached. He *weighed* his parabola to ascertain the area of a segment, and this experiment suggested the theorem that the parabolic area is two-thirds of the area of a circumscribing parallelogram (Figure 9). He admits the value of such experimental methods for arriving at mathematical truths, which afterwards of course must be rigidly proved.

Indeterminate equations, with more unknowns than given equations, have attracted great interest from the earliest days. For example, there may be *one* equation for *two* unknowns:

$$3x - 2y = 5.$$

Many whole numbers x and y satisfy this equation, but it is often interesting to discover the simplest numbers to do so.

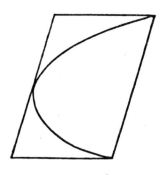

FIGURE 9

Such problems are closely connected with continued fractions, and perhaps Archimedes was beginning to recognize this. At all events we are told that he set the *Cattle Problem* to his friends in Alexandria.

The problem dealt with eight herds, four of bulls and four of cows, according to the colours, white, black, yellow and dappled. Certain facts were stated; for example, that the dappled bulls exceeded the yellow bulls in multitude by $(\frac{1}{6} + \frac{1}{7})$ of the number of white bulls: and the problem required, for its solution, the exact size of each herd. In other words there were eight unknown numbers to be found, but unfortunately, when turned into algebra, the data of the problem provided only seven equations. One of these equations, typical of all seven, can readily be formed from the facts already cited. If x denotes the number of dappled bulls, y that of the white bulls, and z that of the yellow bulls, it follows that

$$x = (\tfrac{1}{6} + \tfrac{1}{7})\, y + z.$$

From seven such equations for eight unknowns, of which only three x, y, z occur in this particular equation, all the unknowns have to be found. There are, of course, an infinite number of solutions to seven equations for eight unknowns. The simplest solution of the above equation, taken apart from its context, is $x = 14$, $y = 42$, $z = 1$. As this does not fit in with the other six equations, a more complicated set of numbers for x, y, z must be found. Who would guess that the smallest value of x satisfying all seven such innocent looking equations is a number exceeding $3\frac{1}{2}$ million? In our decimal notation this is a number seven figures long. But Archimedes improved on the problem by stating that 'when the white bulls joined in number with the black, they stood firm, with depth and breadth of equal measurement; and the plains of Thrinakia, far stretching all ways, were filled with their multitude.' Taking this to mean that the total number of black and white bulls was square, an enterprising investigator, fifty years ago, showed that the smallest such herd amounted to a number 200,000 figures long. The plains of Thrinakia would have to be replaced by the Milky Way.

The so-called *Axiom of Archimedes* bears his name probably because of its application on a grand scale, when he showed that the amount of sand in the world was finite. This appears in the *Sand-Reckoner,* a work full of quaint interest, and important for its influence on the arithmeticians of the last century. The opening sentences run as follows:

'There are some, King Gelon, who think that the number of the sand is infinite in multitude: and I mean by the sand not only that which exists about Syracuse and the rest of Sicily but also that which is found in every region whether inhabited or uninhabited. And again, there are some who, without regarding it as infinite, yet think that no number has been named which is great enough to exceed its multitude.'

So far from weeping to see such quantities of sand, Archimedes cheerfully fancies the whole Universe to be stuffed with sandgrains and then proceeds to count them. After a tilt at

43

the astronomer Aristarchus for talking of the ratio of the centre of a sphere to the surface—'it being easy to see that this is impossible, the centre having no magnitude'—he gently puts Aristarchus right and then turns to the problem. First he settles the question, how many grains of sand placed side by side would measure the diameter of a poppy seed. Then, how many poppy seeds would measure a finger breadth. From poppy seed to finger breadth, from finger breadth to stadium, and so on to a span of 10,000 million stadia, he serenely carries out his arithmetical reductions. Mathematically he is developing something more elaborate than the theory of indices: his arithmetic might be called *the theory of indices of indices,* in which he classifies his gigantic numbers by orders and periods. The first order consists of all numbers from 1 to 100,000,000 $= 10^8$, and the first period ends with the number $10^{800,000,000}$. This number can be expressed more compactly as $(10^8)^{10^8}$, but in the ordinary decimal notation consists of eight hundred million and one figures. Archimedes advances through further periods of this enormous size, never pausing in his task until the hundred-millionth period is reached.

In conclusion:

'I conceive that these things, King Gelon, will appear incredible to the great majority of people who have not studied mathematics, but that to those who are conversant therewith and have given thought to the question of the distances and sizes of the earth, the sun and moon, and the whole universe, the proof will carry conviction. And it was for this reason that I thought the subject would be not inappropriate for your consideration.'

One cannot pass from the story of Archimedes without reference to his work on statics and hydrostatics, in which he created a new application for mathematics. Like the rest of his writings this was masterly. Finally, in a book now lost, he discussed the semiregular solids, which generalize on the Pythagorean group of five regular solids. When each face of the solid is to be a regular polygon, exactly thirteen and no more forms are possible, as Kepler was one of the first to verify.

The third great mathematician of this period was APOLLONIUS of Perga in Pamphilia (? 262–200 B.C.), who earned the title 'the great geometer.' Little is known of him but that he came as a young man to Alexandria, stayed long, travelled elsewhere, and visited Pergamum, where he met Eudemus, one of the early historians of our science. Apollonius wrote extensively, and many of his books are extant. His prefaces are admirable, showing how perfect was the style of the great mathematicians when they were free from the trammels of technical terminology. He speaks with evident pleasure of some results: 'the most and prettiest of these theorems are new.'

What Euclid did for elementary geometry, Apollonius did for conic sections. He defined these curves as sections of a cone standing on a circular base; but the cone may be *oblique*. He noticed that not only were all sections parallel to the base, circular, but that there was also a secondary set of circular sections.

Although a circle is much easier to study than an ellipse, yet every property of a circle gives rise to a corresponding property of an ellipse. For example, if a circle and tangent are looked at obliquely, what the eye sees is an ellipse and its tangent. This matter of perspective leads on to projective geometry; and in this manner Apollonius simplified his problems. By pure geometry he arrived at the properties of conics which we nowadays express by equations such as

$$\frac{x^2}{a^2} \pm \frac{y^2}{b^2} = 1$$

and $ax^2 + bxy + cy^2 = 1$, and even $\sqrt{ax} + \sqrt{by} = 1$. In the second equation a, b, c denote given multiples of certain squares and a rectangle, the total area being constant. From our analytical geometry of conics he had clearly very little to learn except the notation, which improves on his own. He solved the difficult problem of finding the shortest and longest distances from a given point P to a conic. Such lines cut the curve at right angles and are called *normals*. He found that as many as four normals could be drawn from favourable positions of P, and less from others. This led him to consider a still more complicated curve called the *evolute*, which he fully investigated. He worked with what is virtually an equation of the sixth degree in x and y, or its geometrical equivalent—in its day a wonderful feat. His general problem, [*locus*] *ad tres et quattuor lineas*, will be considered when we turn to the work of Pappus.

Another achievement of Apollonius was his complete solution of a problem about a circle satisfying three conditions. When a circle passes through a given point, *or* touches a given line, *or* touches a given circle, it is said to satisfy one condition. So the problem of Apollonius really involved nine cases, ranging from the description of a circle through three given points to that of a circle touching three given circles. The simplest of these cases were probably quite well known: in fact, one of them occurs in the *Elements* of Euclid.

Apollonius was also a competent arithmetician and astronomer. It is reported that he wrote on *Unordered Irrationals*, and invented a 'quick delivery' method of approximating to the number π. Here, to judge from his title, it looks as if he had begun the theory of uniform convergence.

It may now be wondered what was left for their successors to discover after Archimedes and Apollonius had combed the field? So complete was their work that only a few trivial gaps needed to be filled, such as the addition of a focus to a parabola or a directrix to a conic, properties which Apollonius seems to have overlooked. The next great step could not be taken until algebra was abreast of geometry, and until men like Kepler, Cavalieri and Descartes were endowed with both types of technique.

CHAPTER IV

The Second Alexandrian School: Pappus and Diophantus

With the death of Apollonius the golden age of Greek mathematics came to an end. From the time of Thales there had been almost a continuous chain of outstanding mathematicians. But until about the third century A.D., when Hero, Pappus and Diophantus once more brought fame to Alexandria, there seems to have been no mathematician of preeminence. During this interval of about half a millennium the pressure of Roman culture had discouraged Greek mathematics, although a certain interest in mechanics and astronomy was maintained; and the age produced the great astronomer HIPPARCHUS, and two noteworthy commentators, MENELAUS and PTOLEMY. Menelaus lived about the year A.D. 100, and Ptolemy was perhaps fifty years his junior. There is a strange monotony in trying to detail any facts whatsoever about these men—so little is known for certain, beyond their actual writings. The same uncertainty hangs over Hero, Pappus and Diophantus, whose names may be associated together as forming the Second Alexandrian School, because they each appear to have been active about the year A.D. 300. Yet Pappus and Diophantus are surrounded by mystery. Each seems to be a solitary echo of bygone days, in closer touch with Pythagoras and Archimedes than with their contemporaries, or even with each other.

MENELAUS is interesting, more particularly to geometers, because he made a considerable contribution to spherical trigonometry. Many new theorems occur in his writings—new in the sense that no earlier records are known to exist. But it is commonly supposed that most of the results originated with Hipparchus, Apollonius and Euclid. A well-known theorem, dealing with the points in which a straight line drawn across a triangle meets the sides, still bears his name. For some reason, hard to fathom, it is often classed to-day as 'modern geometry,' a description which scarcely does justice to its hoary antiquity. The occasion of its appearance in the work of Menelaus is the more significant because he used it to prove a similar theorem for a triangle drawn on a sphere. Menelaus gave several theorems which hold equally well for triangles and other figures, whether they are drawn on a sphere, or on a flat plane. They include a very fundamental theorem known as the *cross ratio* property of a transversal drawn across a pencil of lines. This too is 'modern geometry.' He also gave the celebrated theorem that the angles of a spherical triangle are together greater than two right angles.

PTOLEMY (? A.D. 100–168), who was a good geometer, will always be remembered for his work in astronomy. He treated this subject with a completeness comparable to that which Euclid achieved in geometry. His compilation is known as the Almagest—a name which is thought to be an Arabic abbreviation of the original Greek title.* His work made a strong appeal to the Arabs, who were attracted by the less abstract branches of mathematics; and through the Arabs it ultimately found a footing in mediaeval Europe. In this way a certain planetary theory called the Ptolemaic system became commonly accepted, holding sway for many centuries until it was superseded by the Copernican system. Ptolemy, following the lead of Hipparchus, chose one of several com-

* Meaning 'The Great Compilation.'

peting explanations of planetary motion, and interpreted the facts by an ingenious combination of circular orbits, or epicycles. Fundamental to his theory was the supposition that the Earth is fixed in space: and, if this is granted, his argument follows very adequately. But there were other explanations, such as that of Aristarchus, the friend of Archimedes, who supposed that the Earth travels round the Sun. When, therefore, Copernicus superseded the Ptolemaic theory by his own well-known system, centred on the Sun, he was restoring a far older theory to its rightful place.

HERO of Alexandria was a very practical genius with considerable mathematical powers. It is generally assumed that all the great mathematicians of the Hellenic world were Greek; but it is supposed that Hero was not. He was probably an Egyptian. At any rate there is in his work a strong bias towards the applications and away from the abstractions of mathematics, which is quite in keeping with the national characteristics of Egypt. Yet Hero proved to be a shrewd follower of Archimedes, bringing his mathematics to bear on engineering and surveying. He not only made discoveries in geometry and physics, but is also reputed to have invented a steam engine. One of his most interesting theorems proves that, when light from an object is broken by reflection on mirrors, the path of the ray between object and eye is a minimum. This is an instance of a *principle of least action,* which was formally adopted for optics and dynamics by Hamilton in the last century, and which has recently been incorporated in the work of Einstein. We may, therefore, regard Hero as the pioneer of Relativity (*c.* A.D. 250).

At the beginning of the fourth century there was a revival of pure mathematics, when something of the Pythagorean enthusiasm for geometry and algebra existed once again in Alexandria under the influence of Pappus and Diophantus. PAPPUS wrote a great commentary called the Collection (συναγωγή); and happily many of his books are preserved. They form a valuable link with still more ancient sources, and particularly with the lost work of Euclid and Apollonius. As an expositor, Pappus rivals Euclid himself, both in completeness of design and wealth of outlook. To discover what Euclid and his followers were about, from reading the Collection, is like trying to follow a masterly game of chess by listening to the comments of an intelligent onlooker who is in full sight of the board.

Pappus was somewhat vain and occasionally unscrupulous, but he had enough sympathy to enter into the spirit of each great epoch. The space-filling figures of Pythagorean geometry made him brood over the marvels of bee-geometry; for God has endowed these sagacious little creatures with a power to construct their honey cells with the *smallest* enclosing surface. How far the bee knows this is not for the mathematician to say, but the fact is perfectly true. Triangular, or square, cells could be crowded together, each holding the same amount of honey as the hexagonal cell; but the hexagonal cell requires least wax. Like the mirrors of Hero, this again suggests *least action* in nature: and Pappus was disclosing another important line of inquiry. He put the question, What is the maximum volume enclosed by a given superficial area? This was perhaps the earliest suggestion of a branch of mathematics called the *calculus of variations*.

Most striking, and in true Archimedean style, is his famous theorem which determines the volume of a surface of revolution. His leading idea may be grasped by first noting that the volume of a straight tube is known if its cross-section A and its length l are given. For the volume is the product A·l. Pappus generalized this elementary result by considering such a tube to be no longer straight but circular. The cross-section

A was taken to be the same at every place; but the length of the tube would need further definition. For example, the length of an inflated bicycle tyre is least if measured round the inner circle in contact with the rim, and is greatest round the outermost circle. This illustration suggests that an *average* or mean length l may exist for which the formula $A \cdot l$ still gives the volume. Pappus found that for such a circular tube this was so, and he located his average length as that of the circle passing through the centroid of each cross-section A. By centroid is meant that special point of a plane area often called the centre of gravity. As the shape of the section A is immaterial to his result, the theorem is one of the most general conclusions in ancient mathematics. In later years P. Guldin (1577–1643) without even the excuse of an unconscious re-discovery, calmly annexed this theorem, and it has become unjustly associated with his name.

As two further examples of important geometrical work by Pappus, the properties of the two following diagrams may be given. There are no hidden subtleties about the drawing of either figure. In the first (Figure 10), A, B, C, D are four points through which various straight lines have been drawn;

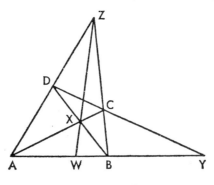

FIGURE 10

and these intersect as shown at X, Y, Z. The line joining ZX is produced and cuts AB at W. The interest of this construction lies in the fact that, no matter what the shape of the quadrilateral ABCD may be, the lines AW, AB, AY are in harmonical progression. In the second figure (Figure 11), ABC and DEF are any two straight lines. These trios of points are joined crosswise by the three pairs of lines meeting at X, Y, Z. Then it follows that X, Y, Z are themselves in line. Here the interest lies in the *symmetry* of the result. It has nine lines meeting by threes in nine points: but it also has nine points lying by threes on the nine lines, as the reader may verify. This nice balance between points and lines of a figure is an early instance of reciprocation, or the *principle of duality,* in geometry.

In the parts of geometry which deal with such figures of points and lines, Pappus excelled. He gave a surprisingly full account of kindred properties connected with the quadrilateral, and particularly with a grouping of six points upon a line into three pairs. This so-called *involution* of six points would be effected by erasing the line ZXW in the first of the above figures, and re-drawing it so as to cross the other six lines at random in six distinct points.

In a significant passage of commentary on Apollonius, Pappus throws light upon what was evidently a very famous problem—the *locus ad tres et quattuor lineas.* It sums up so well the best Greek thought upon conics and it so very nearly inaugurates analytical geometry that it deserves special mention. Apollonius, says Pappus, considered the locus or trace of a roving point P in relation to three or four fixed lines. Suppose P were at a distance x from the first line, y from the second, z from the third, and t from the fourth. Suppose further that these distances were measured in specified directions, but not necessarily at right angles to their several lines. Then,

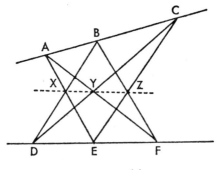

FIGURE 11

as P moves, the values of *x*, *y*, *z*, *t* would vary; although it would always be possible to construct a rectangle of area *xy*, or a solid rectangular block of volume *xyz*. But as space is three-dimensional there is apparently nothing in geometry corresponding to the product *xyzt* derived from four lines. On the other hand the ratio *x* : *y* of two lines is a *number*, and there is nothing to prevent us from multiplying together as many such ratios as we like. So from the four lines *x*, *y*, *z*, *t* we can form two ratios *x* : *y* and *z* : *t*, and then multiply them together. This gives *xz* : *yt*. Now if the resulting ratio is given as a constant, and equal to *c*, we can write

$$xz/yt = c, \text{ or } xz = cyt.$$

This is a way of stating the Apollonian problem about four lines: it indicates that the rectangle of the distances *x*, *z* from P to two of the lines is proportional to that of its distances *y*, *t* from the other two. When this happens, as Apollonius proved, P describes a conic. By a slight modification the same scrutiny may be applied to the problem, if three and not four lines are given. Pappus continues his commentary by generalizing the result with any number of lines: but it will be clearest if we confine ourselves to six lines.

54

If the distances of the point P from six given lines are x, y, z, u, v, w, then we can form them into three ratios $x : y$, $z : u$, $v : w$. If, further, it is given that the product of these three ratios is fixed, then we can write

$$\frac{x}{y} \cdot \frac{z}{u} \cdot \frac{v}{w} = c.$$

Pappus draws the correct conclusion that, when this happens, the point P is constrained to lie upon a certain locus or curve. But after a few more remarks he turns aside as if ashamed of having said something obvious. He had nevertheless again made one of the most general statements in all ancient geometry. He had begun the theory of Higher Plane Curves. For the number of ratios involved in such a constant product defines what is called the *order* of the locus. So a conic is a curve of order two, because it involves two ratios, as is shown in the Apollonian case above. In the simpler case, when only one ratio $x : y$ is utilized, the locus is a straight line. For this reason a straight line is sometimes called a curve of the first order. But Pappus had suggested curves of order higher than the second. These are now called cubics, quartics, quintics, and so on. To be sure, particular cases of cubic and other curves had already been discovered. The ancients had invented them for *ad hoc* purposes of trisecting an angle, and the like: but mathematicians had to wait for Descartes to clinch the matter.

The other great mathematician who brought fame to Alexandria was DIOPHANTUS. He is celebrated for his writings on algebra, and lived at the time of Pappus, or perhaps a little earlier. This we gather from a letter of Psellus, who records that Anatolius, Bishop of Laodicea about A.D. 280, dedicated to Diophantus a concise treatise on the Egyptian method of reckoning. Diophantus was devoted to algebra, as the wording of a Greek epigram indicates, which tells us the scanty record of his life. His boyhood lasted ⅙th of his life; his beard grew after ¹⁄₁₂th more; he married after ⅐th more, and his son was born five years later; the son lived to half his father's age, and the father died four years after his son.

If x was the age when he died, then,

$$\tfrac{1}{6}x + \tfrac{1}{12}x + \tfrac{1}{7}x + 5 + \tfrac{1}{2}x + 4 = x;$$

and Diophantus must have lived to be eighty-four years old.

The chief surviving writings of Diophantus are six of the thirteen books forming the *Arithmetica,* and fragments of his *Polygonal Numbers* and *Porisms.* Twelve hundred years after they were written these books began to attract the attention of scholars in Europe. As Regiomontanus observed in 1463: 'In these old books the very flower of the whole of arithmetic lies hid, the *ars rei et census* which to-day we call by the Arabic name of Algebra.' This work of Diophantus has a two-fold importance: he made an essential improvement in mathematical notation, while at the same time he added large instalments to the scope of algebra as it then existed. The full significance of his services to mathematics only became evident with the rise of the early French school in the fifteenth and sixteenth centuries.

The study of notation is interesting, and covers a wider sphere than at first sight may be supposed. For it is the study of symbols; and as words are symbols of thought, it embraces literature itself. Now we may concentrate our attention on the literal symbol as it appears to the eye in a mathematical formula and in a printed sentence; or else on the thing signified, on the sense of the passage, and on the thought behind the symbol. A good notation is a valuable tool; it brings its own fitness and suggestiveness, it is easy to recognize and is comfortable to use. Given this tool and the material to work upon, advance may be expected. In their own language and in their geometrical notation the Greeks were well favoured: and a due succession of triumphs followed. But their arithmetic and algebra only advanced in spite of an unfortunate notation. For the Greeks were hampered by their use of letters α, β, γ for the numbers 1, 2, 3, and this concealed from them the flexibility of ordinary arithmetical calculations. On the other hand, the very excellence of our *decimal notation* has made these operations well-nigh trivial. Before the notation was widely known, even simple addition, without the help of a ball frame, was a task of some skill. The chief merits of this notation are the sign 0 for zero, and the use of one symbol, its meaning being decided by its context, to denote several distinct things, as, for example, the writing of 11 to denote *ten* and *one*. The history of this usage has been traced to a source in Southern India, dating shortly after the time of Diophantus. Thence it spread to the Moslem world and so to mediaeval Europe.

In the previous chapters many algebraic formulae have occurred. They are, of course, not a literal transcription of the Greek, but are concise symbolic statements of Greek theorems originally given in verbal sentences or in geometrical form. For instance, a^2 has been used instead of 'the square on AB.' The earliest examples of this symbolic algebra occur in the work of Vieta, who lived in the sixteenth century, though it only came into general use about the year A.D. 1650. Until that time the notation of Diophantus had been universally adopted.

An old classification speaks of

Rhetorical Algebra,
Syncopated Algebra,
Symbolic Algebra,

and these names serve to indicate broad lines of development. By the rhetorical is meant algebra expressed in ordinary language. Then syncopations, or abbreviations, similar to our use of H.M.S. for His Majesty's Ship, and the like, became common among the ancients. To Diophantus more than to any other we owe this essential improvement. The third, symbolic algebra, became finally established, once Vieta had invented it, through the influence of Napier, Descartes and Wallis.

A typical expression of symbolic algebra is

$$(250x^2 + 2520) \div (x^4 + 900 - 60x^2):$$

and this serves to indicate the type of complication which Diophantus successfully faced. His syncopations enabled him to write down, and deal with, equations involving this or similar expressions. For $250x^2$ he wrote Δγσν: here the letter ν meant 50 and σ, 200, according to the ordinary Greek practice. But the Δγ was short for the Greek word meaning *power* (it is our word, dynamic); and power meant the square of the unknown number. Diophantus used the letter ς for the first power of the unknown, and the abbreviation of the word *cube* for the third power. He used no sign for *plus*, but a sort of inverted ψ for *minus*, the letter ι for *equals*, and a special phrase to denote the *division* of one expression by another. It is interesting that his idea of addition and subtraction was 'forthcoming' and 'wanting,' and that the Greek word for wanting is related to the Pythagorean term *ellipse*.

Those who have solved quadratic equations will remember the little refrain—'the square of half the co-efficient of *x*.' It is a quotation from Diophantus, who dealt with such equations very thoroughly. He even ventured on the easier cases of cubic equations. Yet he speaks of 'the impossible solution of the absurd equation $4 = 4x + 20$': such an equation requires a negative solution; and it was not until much later that negative numbers as things in themselves were contemplated. But fractions and alternative roots of quadratic equations presented to him no difficulties.

We need not go far into the puzzles of 'problems leading to simple equations' to convince ourselves of the value of using several letters x, y, z for the unknown quantities. Each different symbol comes like a friendly hand to help in disentangling the skein. As Diophantus attempted such problems with the sole use of his symbol ς he was, so to speak, tying one hand behind his back and successfully working single-handed. This was clearly the chief drawback of his notation. Nevertheless he cleverly solved simultaneous equations such as

$$yz = m\,(y + z),\ zx = n\,(z + x),\ xy = p\,(x + y);$$

and it is evident from this instance that he saw the value of *symmetry* in algebra.

All this is valuable for its general influence upon mathematical manipulation: and had the genius of Diophantus taken him no farther, he would still be respected as a competent algebraist. But he attained far greater heights, and his abiding work lies in the Theory of Numbers and of Indeterminate Equations. Examples of these last occurred in the *Cattle Problem* of Archimedes (p. 42) and in the relation $2x^2 - y^2 = 1$ (of p. 29). His name is still attached to simple equations, such as enter the Cattle Problem, although he never appears to have interested himself in them. Instead he was concerned with the more difficult quadratic and higher types, as, for example, the equation

$$x^4 + y^4 + z^4 = u^2.$$

He discovered four whole numbers x, y, z, u for which this statement was true. Centuries later his pages were eagerly read by Fermat, who proved to be a belated but brilliant disciple. 'Why,' says Fermat, 'did not Diophantus seek *two* fourth powers such that their sum is square? This problem is, in fact, impossible, as by my method I am able to prove with all rigour.' No doubt Diophantus had experimented far enough with the easier looking equation $x^4 + y^4 = u^2$ to prove that no solution was available.

This brings us to the close of the Hellenic period; and we are now in a position to appreciate the contribution which the Greeks made to mathematics. They virtually sketched the whole design that was to give incessant opportunities for the mathematicians and physicists of later centuries. In some parts of geometry and in the theory of the irrational the picture had been actually completed.

Within a glittering heap of numerical and geometrical puzzles and trifles—the accumulation in Egypt or the East of bygone ages—the Greeks had found order. Their genius had made mathematics and music out of the discord. And now in turn their own work was to appear as a wealth of scattered problems whose interrelations would be seen as parts of a still grander whole. New instruments were to be invented— the decimal notation, the logarithm, the analytical geometry of Descartes, and the magnifying glass. Each in its way has profoundly modified and enriched the mathematics handed down by the Greeks. Such profound changes have been wrought that we have been in some danger of losing a proper perspective in mathematics as a whole. So ingrained to-day is our habit of microscopic scrutiny that we are apt to think that all accuracy is effected by examining the infinitesimal under a glass or by reducing everything to decimals. It is well to remember that, even in the scientific world, this is but a partial method of arriving at exact results. Speaking numerically, multiplication, and not division, was the guiding process of the Greeks. The spacious definition of equal ratios which the astronomer Eudoxus bequeathed was not the work of a man with one eye glued to a micrometer.

CHAPTER V

The Renaissance:
Napier and Kepler;
The Rise of Analysis

After the death of Pappus, Greek mathematics and indeed European mathematics lay dormant for about a thousand years. The history of the science passed almost entirely to India and Arabia; and by far the most important event of this long period was the introduction of the Indian decimal notation into Europe. The credit for this innovation is due to Leonardo of Pisa, who was mentioned on p. 30, and certainly ranks as a remarkable mathematician in these barren centuries. From time to time there were others of merit and even of genius; but, judged by the lofty standard of past achievements and of what the future held in store, no one rose supreme. The broad fact remains: Pappus died in the middle of the fourth century, and the next great forward step for Western mathematics was taken in the sixteenth century.

It is still an obscure historical problem to determine whether Indian mathematics is independent of Greek influence. When Alexander conquered Eastern lands he certainly reached India, so that at any rate there was contact between East and West. This took place about 300 B.C., whereas the early mathematical work of India is chiefly attributed to the far later period A.D. 450–650. So in the present state of our knowledge it is safest to assume that considerable independent work was done in India. An unnamed genius invented the decimal notation; he was followed by ARYABHATA and BRAHMAGUPTA, who made substantial progress in algebra and trigonometry. Their work brings us to the seventh century, an era marked by the fall of Alexandria and the rise of the Moslem civilization.

The very word *algebra* is part of an Arabic phrase for 'the science of reduction and cancellation,' and the digits we habitually use are often called the Arabic notation. These survivals remind us that mathematical knowledge was mediated to western Europe through the Arabs. But it will be clear from what has already been said that the Arabs were in no sense the originators of either algebra or the number notation. The Arabs rendered homage to mathematics; they valued the ancient learning whether it came from Greece or India. They proved apt scholars; and soon they were industriously translating into Arabic such valuable old manuscripts as their forerunners had not destroyed. In practical computation and the making of tables they showed their skill, but they lacked the originality and genius of Greece and India. Great tracts of Diophantine algebra and of geometry left them quite unmoved. For long centuries they were the safe custodians of mathematical science.

Then came the next chapter in the story, when northern Italy and the nations beyond the Alps began to feel their wakening strength. Heart and mind alike were stirred by the great intellectual and spiritual movements of the Renaissance and the Reformation. Once again mathematics was investigated with something of the ancient keenness, and its study was greatly stimulated by the invention of printing. There were centres of learning, in touch with the thriving city life of Venice and Bologna and other celebrated towns of mediaeval Europe. Italy led the way; France, Scotland, Germany and England were soon to follow. The first essential advance beyond Greek and Oriental mathematics was made by SCIPIO FERRO (1465–1526), who picked up the threads where

Diophantus left them. Ferro discovered a solution to the cubic equation:

$$x^3 + mx = n;$$

and, as this solved a problem that had baffled the Greeks, it was a remarkable achievement.

Scipio was the son of a paper-maker in Bologna whose house can still be precisely located. He became Reader in mathematics at the University in 1496 and continued in office, except for a few years' interval at Venice, till his death in the year 1526. In those days mathematical discoveries were treasured as family secrets, only to be divulged to a few intimate disciples. So for thirty years this solution was carefully guarded, and it only finally came to light owing to a scientific dispute. Such wranglings were very fashionable: they were the jousts and tournaments of the intellectual world, and mathematical devices, often double-edged, were the weapons. Some protagonists preferred to spar with slighter blades—only drawing their mightiest swords as a last resort. Among them were Tartaglia and Cardan, both very celebrated, and ranking with Scipio as leading figures in this drama of the Cubic Equation. Scipio himself was dragged rather unwillingly into the fray: others relished it.

Niccolo Fontana (1500–1557) received the nickname Tartaglia because he stammered. When he was quite a little lad he had been almost killed by a wound on the head, which permanently affected his speech. This had occurred in the butchery that followed the capture of Brescia, his native town, by the French. His father, a postal messenger, was amongst the slain, but his mother escaped, and rescued the boy. Although they lived in great poverty, Tartaglia was determined to learn. Lacking the ordinary writing materials, he even used tombstones as slates, and eventually rose to a position of eminence for his undoubted mathematical ability. He emulated Ferro by solving a new type of cubic equation, $x^3 + mx^2 = n;$ and when he heard of the original problem, he was led to re-discover Ferro's solution. This is an interesting example of what frequently happens,—the mere knowledge that a certain step had been taken being inducement enough for another to take the same step. Tartaglia was the first to apply mathematics to military problems in artillery.

GIROLAMO CARDAN (1501–1576) was a turbulent man of genius, very unscrupulous, very indiscreet, but of commanding mathematical ability. With strange versatility he was astrologer and philosopher, gambler and algebraist, physician yet father and defender of a murderer, heretic yet receiver of a pension from the Pope. He occupied the Chair of Mathematics at Milan and also practiced medicine. In 1552 he visited Scotland at the invitation of John Hamilton, Archbishop of St. Andrews, whom he cured of asthma. He was interested one day to find that Tartaglia held a solution of the cubic equation. Cardan begged to be told the details, and eventually under a pledge of secrecy obtained what he wanted. Then he calmly proceeded to publish it as his own unaided work in the *Ars Magna,* which appeared in 1545. Such a blot on his pages is deplorable because of the admittedly original algebra to be found in the book. He seems to have been equally ungenerous in the treatment of his pupil Ferrari, who was the first to solve a quartic equation. Yet Cardan combined piracy with a measure of honest toil; and he had enough mathematical genius in him to profit by these spoils. He opened up the general theory of the cubic and quartic equations, by discussing how many roots an equation may have. He surmised the need not only for negative but for complex (or imaginary) numbers to effect complete solutions. He also found out the more important relations between the roots.

By these mathematical achievements, so variously conducted, Italy made a substantial advance. It was now possible to state, in an algebraic formula, the solution of the equation

$$ax^4 + bx^3 + cx^2 + dx + e = 0.$$

The matter had proceeded step by step from the simple to the quadratic, the cubic and the quartic equation. Naturally the question of the quintic and higher equations arose, but centuries passed before further light was thrown upon them. About a hundred years ago a young Scandinavian mathematician named Abel found out the truth about these equations. They proved to be insoluble by finite algebraic formulae such as these Italians had used. Cardan, it would seem, had unwittingly brought the algebraic theory of equations to a violent full stop!

Now what was going on at this time elsewhere in Europe? Something very significant in Germany, and a steady preparation for the new learning in France, Flanders and England. Contemporary with Scipio Ferro were three German pioneers, DÜRER, STIFEL and COPERNICUS. Dürer is renowned for his art; Stifel was a considerable writer on algebra; and Copernicus revolutionized astronomy by postulating that the Earth and all the planets revolve around the Sun as centre. About this time, in 1522, the first book on Arithmetic was published in England: it was a fine scholarly production by TONSTALL, who became Bishop of London. In the preface the author explains the reason for his belated interest in arithmetic. Having forgotten what he had learnt as a boy, he realized his disadvantage when certain gold- and silver-smiths tried to cheat him, and he wished to check their transactions.

Half a century later another branch of mathematics came into prominence, when STEVINUS left his mark in work on

Statics and Hydrostatics. He was born at Bruges in 1548, and lived in the Low Countries. Then once more the scene shifts to Italy, where GALILEO of Pisa (1564–1642) invented dynamics, by rebuilding the scanty and ill-conceived system which had come down from the time of Aristotle. Galileo showed the importance of experimental evidence as an essential prelude to a theoretical account of moving objects. This was the beginning of physical science—which really lies outside our present scope—and by taking this step Galileo considerably enlarged the possible applications of mathematics. In such applications it was no longer possible for the mathematician to make his discoveries merely by sitting in his study or by taking a walk. He had to face stubborn facts, often very baffling to common sense, but always the outcome of systematic experiments. Two of the first to do this were Galileo and his contemporary, Kepler. Galileo found out the facts of dynamics for himself by dropping pebbles from a leaning tower at Pisa. Kepler took, for the basis of his astronomical speculations, the results of patient observations made by Tycho Brahe, of whom more anon.

The latter half of the sixteenth century also saw the rise of mathematics in France and Scotland. France produced VIETA, and Scotland NAPIER. The work of these two great men reminds us how deep was the influence of Ancient Greece upon the leaders of this mathematical Renaissance. Allusion has already been made to the share which Vieta took in improving the notation of algebra: he also attacked several outstanding problems that had baffled the Greeks, and he made excellent progress. He showed, for example, that the famous problem of trisecting an angle really depended on the solution of a cubic equation. Also he reduced the problem of squaring a circle to that of evaluating the elegant expression:

$$\frac{2}{\pi} = \sqrt{½} \times \sqrt{(½ + \sqrt{½})} \times ½ \sqrt{(½ + ½\sqrt{(½ + ½\sqrt{½})})} \times \cdots$$

Here was a considerable novelty—the first actual formula for the time-honoured number π, which Archimedes had located to lie somewhere between $3\frac{1}{7}$ and $3\frac{10}{71}$. Vieta was also the first to make explicit use of that wonderful principle of duality, or reciprocation, which was hinted at by Pappus. We had an instance in the figure 11 of p. 54. For Vieta pointed out the importance of a polar triangle, obtained from a spherical triangle ABC. He drew three great circular arcs whose poles were respectively A, B, C; and then he formed a second triangle from these arcs. The study of the two triangles jointly turned out to be easier than that of the original triangle by itself.

Perhaps the most remarkable of all these eminent mathematicians was JOHN NAPIER, Baron of Merchiston, who discovered the logarithm. This achievement broke entirely new ground, and it had great consequences, both practical and theoretical. It gave not only a wonderful labour-saving device for arithmetical computation, but it also suggested several leading principles in higher analysis.

John Napier was born in 1550 and died in 1617: he belonged to a noble Scottish family notable for several famous soldiers. His mother was sister of Adam Bothwell, first reformed Bishop of Orkney, who assisted at the marriage of his notorious kinsman, the Earl of Bothwell, to Queen Mary, and who also anointed and crowned the infant King James VI. Scotland was a country where barbarous hospitality, hunting, the military art and keen religious controversy occupied the time and attention of Napier's contemporaries: a country of baronial leaders whose knowledge of arithmetic went little farther than counting on the fingers of their mail-clad hands.

It was a strange place for the nurture of this fair spirit who seemed to belong to another world. The boy lost his mother when he was thirteen, and in the same year was sent to the University of St. Andrews, where he matriculated in 'the triumphant college of St. Salvator.' In those days St. Andrews was no home of quiet academic studies: accordingly the Bishop, who always took a kindly interest in the lad, advised a change. 'I pray you, schir,' he wrote to John's father, 'to send your son Jhone to the schuyllis; oyer to France or Flanderis; for he can leyr na guid at hame, nor get na proffeit in this maist perullous worlde.' So abroad he went; but it is probable that he soon returned to Merchiston, his home near Edinburgh, where he was to spend so many years of his serene life.

During the year at St. Andrews his interest was aroused in both arithmetic and theology. The preface to his *Plain Discovery of the Whole Revelation of St. John,* which was published in 1593, contains a reference to his 'tender yeares and barneage in Sanct Androis' where he first was led to devote his talents to the study of the Apocalypse. His book is full of profound but, it is to be feared, fruitless speculations; yet in form it follows the finest examples of Greek mathematical argument, of which he was master, while in sober manner of interpretation it was far ahead of its time. Unlike Cardan, before him, and Kepler, after him, he was innocent of magic and astrology.

Napier acquired a great reputation as an inventor; for with his intellectual gifts he combined a fertile nimbleness in making machines. His constant efforts to fashion easier modes of arithmetical calculation led him to produce a variety of devices. One was a sort of chess-arithmetic where digits moved like rooks and bishops on a board: another survives under the name of Napier's Bones. But what impressed his friends was a piece of artillery of such appalling efficiency that it was able to

kill all cattle within the radius of a mile. Napier, horrified, refused to develop this terrifying invention, and it was forgotten.

During his sojourn abroad he eagerly studied the history of
the Arabic notation, which he traced to its Indian source. He
brooded over the mysteries of arithmetic and in particular
over the principle which underlies the number notation. He
was interested in reckoning not only, as is customary, in *tens,*
but also in *twos.* If the number eleven is written 11, the notation indicates *one* ten and *one.* In the common scale of ten
each number is denoted by so many *ones,* so many *tens,* so
many *hundreds,* and so on. But Napier also saw the value
of a binary scale—in which a number is broken up into parts
1, 2, 4, 8, etc. Thus he speaks with interest of the fact that
any number of pounds can be weighed by loading the other
scale pan with one or more from among the weights 1 lb., 2
lb., 4 lb., 8 lb., and so on.

When Napier returned to Scotland he wrote down his
thoughts on arithmetic and algebra, and many of his writings
remain. They are very systematic, showing a curious mixture
of theory and practice: the main business is the theory, but
now and then comes an illustration that 'would please the
mechanicians more than the mathematicians.' Somewhere on
his pages the following table appears:

	I	II	III	IIII	V	VI	VII . . .
1	2	4	8	16	32	64	128 . . .

Perhaps the reader thinks that it is simple and obvious; yet
in the light of the sequel, it is highly significant. Men were
still feeling for a notation of indices, and the full implications of the Arabic decimal notation had hardly yet been
grasped. Napier was looking with the eyes of a Greek-trained
mathematician upon this notation as upon a new plaything.
He saw in the above parallel series of numbers the matching

of an arithmetical with a geometrical progression. A happy inspiration made him think of these two progressions as *growing continuously from term to term*. The above table then became to him a sort of slow kinematograph record, implying that things are happening *between* the recorded terms. By the year 1590, or perhaps earlier, he discovered logarithms—the device which replaces multiplication by addition in arithmetic: and his treatment of the matter shows intimate knowledge of the correspondence between arithmetical and geometrical progressions. So clearly did he foresee the practical benefit of logarithms in astronomy and trigonometry, that he deliberately turned aside from his speculations in algebra, and quietly set himself the lifelong task of producing the requisite tables. Twenty-five years later they were published.

Long before the tables appeared, they created a stir abroad. There dwelt on an island of Denmark the famous Tycho Brahe, who reigned in great pomp over his sea-girt domain. It was called Uraniburg—the Castle of the Heavens—and had been given to him by a beneficent monarch, King Frederick II, for the sole purpose of studying astronomy. Here prolonged gazings and much accurate star chronicling proceeded; but the stars in their courses were getting too much for Tycho. Like a voice from another world word came of a portentous arithmetical discovery in Scotland, the *terra incognita*. The Danish astronomer looked for an early publication of the logarithmic tables; but it was long before they were completed. Napier, in fact, was slow but sure. 'Nothing,' said he, 'is perfect at birth. I await the judgment and criticism of the learned on this, before unadvisedly publishing the others and exposing them to the detraction of the envious.' The first tables appeared in 1614, and immediately attracted the attention of mathematicians in England and on the Continent—notably BRIGGS and KEPLER. The friendship between Napier and Briggs rapidly grew, but was very soon to be cut short: for in 1617, worn out by his incessant toil, Napier died. One of his last writings records how 'owing to our bodily weakness we leave the actual computation of the new canon to others skilled in this kind of work, more particularly to that very learned scholar, my dear friend, Henry Briggs, public Professor of Geometry in London.'

A picturesque account of their first meeting has been handed down. The original publication had so delighted Briggs that

'he could have no quietness in himself, until he had seen that noble person whose only invention they were. . . . Mr. Briggs appoints a certain day when to meet in Edinburgh; but failing thereof, Merchiston was fearful he would not come. It happened one day as John Marr and the Lord Napier were speaking of Mr. Briggs: "Ah, John," saith Merchiston, "Mr. Briggs will not now come": at the very instant one knocks at the gate; John Marr hasted down and it proved to be Mr. Briggs, to his great contentment. He brings Mr. Briggs up into My Lord's chamber, where almost one quarter of an hour was spent, each beholding other with admiration before one word was spoken: at last Mr. Briggs began. "My Lord, I have undertaken this long journey purposely to see your person, and to know by what engine of wit or ingenuity you came first to think of this most excellent help unto Astronomy, viz. the Logarithms: but My Lord, being by you found out, I wonder nobody else found it out before, when now being known it appears so easy." '

Exactly: and perhaps this was the highest praise. It is pleasant to record the excellent harmony existing between Napier, Briggs and Kepler. Kepler the same year discovered his third great planetary *canon* which he published in the *Ephemerides* of 1620, a work inscribed to Napier; and there for frontispiece was a telescope of Galileo, the elliptic orbit of a planet, the system of Copernicus, and a female figure with the Napierian logarithm of half the radius of a circle arranged as a glory round her head!

And what was a logarithm? Put into unofficial language it can be explained somewhat as follows. A point G may be conceived as describing a straight line TS with diminishing speed, slowing towards its destination S, in such wise that the speed is always proportional to the distance it has to go. When the point G is at the place *d* its speed is proportional to the distance *d*S. What a problem in dynamics to launch on the

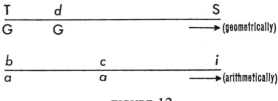

FIGURE 12

world, before dynamics were even invented! This motion Napier called *decreasing geometrically*. Alongside this, and upon a parallel line *bi,* a point *a* moves off uniformly from its starting position *b*. This Napier called *increasing arithmetically*. The race between the moving points G and *a* is supposed to begin at T and *b,* both starting off at the same speed; and then at any subsequent instant the places reached by G and *a* are recorded. When G has reached *d* let *a* have reached *c*. Then the number measuring the length *bc* is called by Napier the *logarithm* of the number measuring *d*S. In short, the distance *a* has gone is the logarithm of the distance G has to go.

Beginning with this as his definition Napier built up not only the theoretical properties of logarithms, but also his seven-figure tables. The definition is in effect the statement of a differential equation; and his superstructure provides the complete solution. It even suggests a theory of functions on a genuinely arithmetical basis. As this was done before either the theory of indices or the differential calculus had been invented, it was a wonderful performance.

Napier was also a geometer of some imagination. He devised new methods in spherical trigonometry. Particularly beautiful is his treatment of a right-angled spherical triangle as part of a fivefold figure, reminiscent of the Pythagorean symbol.

The story of Napier shows how the time was ripe for logarithms to be invented, and it is scarcely surprising that another should also have discovered them. This was his contemporary Bürgi, a Swiss watchmaker, who reached his conclusions through the idea of indices, and published his results in 1620. Great credit must also be given to Briggs for the rapid progress he made in fashioning logarithmic tables of all kinds. None but an expert mathematician of considerable originality could have done the work so quickly.

The rapid spread of Napier's logarithms on the Continent was due to the enthusiasm of Kepler, an astronomer, who was born in 1571 of humble parents near Stuttgart in Würtemberg, and died at Ratisbon in 1630. He was a man of affectionate disposition, abundant energy and methodical habits, with the intuition of true genius and the readiness to look for new relations between familiar things. He combined a love of general principles with the habit of attending to details. To his knowledge of ancient and mediaeval lore which included, in one comprehensive grasp, the finest Greek-geometry and the extravagances of astrology, he added the new learning of Copernicus and Napier. He learnt of the former in his student days at Tübingen whence at the age of twenty-two he migrated to Gratz in Austria, where he was appointed Professor.

73

There he imprudently married a wealthy widow—a step which brought him no happiness. Within three years of his appointment he became famous through the publication of his *Mysterium,* a work full of fancies and strange theories of the heavens.

Kepler's interest in the stars and planets developed as he corresponded with the great Tycho Brahe at Uraniburg, who held even kings spellbound by his discoveries. When in course of time Brahe lost royal favour and began to wander, he accepted a post at the new observatory near Prague. He even persuaded Kepler, who also was rather unsettled, to become his assistant. This arrangement was made in 1599 at the instigation of Rudolph II, a taciturn monarch much addicted to astrology, who hoped that these two astrological adepts would bring distinction to his kingdom. In this he was disappointed: for collaboration was not a success between these two strong personalities, with their widely different upbringing. Yet the experience was good for Kepler, especially as he also came under the influence of Galileo. It helped to stabilize his wayward genius. When Tycho died in 1601, Kepler succeeded him as astronomer; but his career was dogged by bad luck. He was often unpaid; his wife died;—nor did a second matrimonial venture prove more successful, although he acted with the greatest deliberation: for he carefully analysed and weighed the virtues and defects of several young ladies until he found his desire. It is a warning to all scientists that there *are* matters in life which elude weights and measures. The axiom of Archimedes has its limitations!

Kepler brimmed over with new ideas. Possessed with a feeling for number and music, and imbued through and through with the notions of Pythagoras, he sought for the underlying harmony in the cosmos. Temperamentally he was as ready to listen as to look for a clue to these secrets. Nor was there any current scientific reason to suppose that light would yield more significant results than sound. So he brought all his genius to bear on the problem of the starry universe: and he dreamt of a harmony in arithmetic, geometry and music that would solve its deepest mysteries. Eventually he was able to disclose his great laws of planetary motion, two in 1609, and the third and finest in the *Harmonices Mundi* of 1619.

These laws, which mark an epoch in the history of mathematical science, are as follows:

1. The orbit of each planet is an ellipse, with the sun at a focus.

2. The line joining the planet to the sun sweeps out equal areas in equal times.

3. The square of the period of the planet is proportional to the cube of its mean distance from the sun.

The period in the case of the earth is, of course, a year. So this third law states that a planet situated twice as far from the sun would take nearly three years to perform its orbit, since the cube of two is only a little less than the square of three. This first law itself made a profound change in the scientific outlook upon nature. From ancient times until the days of Copernicus and Tycho Brahe, circular motion had reigned supreme. But the circle was now replaced by the ellipse: and with the discovery that the ellipse was a path actually performed in the heavens and by the earth itself, a beautiful chapter in ancient geometry had unexpectedly become the centre of a practical natural philosophy. In reaching this spectacular result Kepler inevitably pointed out details in the abstract theory that Apollonius had somehow missed—such as the importance of the focus of a conic, and even the existence of a focus for a parabola. Then by a shrewd combination of his new ideas with the original conical properties, Kepler began to see ellipses, parabolas, hyperbolas, circles, and pairs of lines as so many phases of *one* type of curve. To Kepler, starlight, radiating from points unnumbered leagues away, suggested that in geometry parallel lines have a common point at infinity. Kepler therefore not only found out something to interest the astronomer; he made essential progress in geometry. An enthusiastic geometer once lamented that here was a genius spoilt for mathematics by his interest in astronomy!

The second law of Kepler is remarkable as an early example of the infinitesimal calculus. It belongs to the same order of mathematics as the definition that Napier gave for a logarithm. Again we must remember that this calculus, as a formal branch of mathematics, still lay hidden in the future. Yet Kepler made further important contributions by his accurate methods of calculating the size of areas within curved boundaries. His interest in these matters arose partly through reading the ancient work of Archimedes and partly through a wish to improve on the current method of measuring wine casks. Kepler recorded his results in a curious document, which incidentally contained an ingenious number notation based on the Roman system, where subtraction as well as addition is involved. Kepler used symbols analogous to I, V, X, L, but instead of the numbers one, five, ten and fifty he selected one, three, nine, twenty-seven, and so on. In this way he expressed any whole number very economically; for instance,

$$20 = 27 - 9 + 3 - 1.$$

As an algebraist he also touched upon the theory of recurring series and difference relations. He performed prodigies of calculation from the sheer love of handling numbers. The third of his planetary laws, which followed ten years after the other two, was no easy flight of genius: it represented prolonged hard work.

76

Something may be quoted of the contents in the *Harmonices Mundi* which enshrines this great planetary law. It is typical of the work of this extraordinary man. In it he makes a systematic search into the theory of musical intervals, and their relations to the distances between the planets and the sun: he discusses the significance of the five Platonic regular solids for interplanetary space: he elaborates the properties of the thirteen semi-regular solids of Archimedes: he philosophizes on the place of harmonic and other algebraic progressions in civil life, drawing his illustrations from the dress of Cyrus as a small boy, and the equity of Roman marriage laws. Few indeed are the great discoverers in science who can rival Kepler in richness of imagery! For Kepler, every planet sang its tune: Venus a monotone, the Earth (in the sol-fa notation) the notes *m, f, m,* signifying that in this world man may expect but misery and hunger. This gave Kepler an opportunity for a Latin pun—'in hoc nostro domicilio *mi*seriam et *fa*men obtinere.' The italics are his, and in fact the whole book was written in solemn mediaeval Latin. The song of Mercury, in his arpeggio-like orbit, is

$$d\ r\ m\ f\ s\ l\ t\ d'\ r'\ m'\ d'\ s\ m\ d$$

—stated originally of course in the staff notation. As for the comets, surely they must be live things, darting about with will and purpose 'like fishes in the sea'! This frisky skirl of Mercury amid the sober hummings of the other planets, is no idle fancy: it duly records a curious fact, that the orbit of Mercury is more strongly elliptical, and less like a circle, than that of any other planet. It was this very peculiarity of Mercury which provided Einstein with one of his clues leading to the hypothesis of Relativity.

Carlyle, in his *Frederick the Great* (Book III, Chapter XIV) has preserved a delightful picture of John Kepler as he appeared to a contemporary, Sir Henry Wotton, Ambassador to the King of Bohemia.

" 'He hath a little black Tent . . . ,' says the Ambassador, 'which he can suddenly set up where he will in a Field; and it is convertible (like a windmill) to all quarters at pleasure; capable of not much more than one man, as I conceive, and perhaps at no great ease; exactly close and dark,—save at one hole, about an inch and a half in the diameter to which he applies a long perspective Trunk, with the convex glass fitted to the said hole, and the concave taken out at the other end . . .' . . . An ingenious person, truly, if there ever was one among Adam's Posterity. Just turned fifty, and ill-off for cash. This glimpse of him, in his little black tent with perspective glasses, while the Thirty-Years War blazes out, is welcome as a date."

CHAPTER VI

Descartes and Pascal: The Early French Geometers and their Contemporaries

Hitherto the mathematicians of outstanding ability, whose names have survived, have been comparatively few; but from the beginning of the seventeenth century the number increased so rapidly that it is quite impossible in a short survey to do justice to all. In France alone there were as many mathematicians of genius as Europe had produced during the preceding millennium. Three names will therefore be singled out to be representatives of their time, Descartes and Pascal from among the French, and Newton from among the English. In this heroic age that followed the performances of Napier and Kepler, mathematics attained a remarkable prestige. The age was mathematical; the habits of mind were mathematical; and its methods were deemed necessary for an exact philosophy, or an exact anything else. It was the era when what is called modern philosophy began; and the pioneers among its philosophers, like the Greek philosophers of old, were expert mathematicians. They were Descartes and Leibniz.

DESCARTES was born of Breton parents near Tours in 1596 and died at Stockholm in 1650. In his youth he was delicate, and until the age of twenty his friends despaired of his life. After receiving the traditional scholastic education of mathematics, physics, logic, rhetoric and ancient languages, at which he was an apt pupil, he declared that he had derived no other benefit from his studies than the conviction of his utter ignorance and profound contempt for the systems of philosophy then in vogue.

'And this is why, as soon as my age permitted me to quit my preceptors,' he says, 'I entirely gave up the study of letters; and resolving to seek no other science than that which I could find in myself or else in the great book of the world, I employed the remainder of my youth in travel, in seeing courts and camps, in frequenting people of diverse humours and conditions, . . . and above all in endeavouring to draw profitable reflection from what I saw. For it seemed to me that I should meet with more truth in the reasonings which each man makes in his own affairs, and which if wrong would be speedily punished by failure, than in those reasonings which the philosopher makes in his study.'

In this frame of mind he led a roving, unsettled life; sometimes serving in the army, sometimes remaining in solitude. At the age of three and twenty, when residing in his winter quarters at Neuberg on the Danube, he conceived the idea of a reformation in philosophy. Thereupon he began his travels, and ten years later retired to Holland to arrange his thoughts into a considered whole. In 1638 he published his *Discourse on Method* and his *Meditations*. An immense sensation was produced by the *Discourse*, which contained important mathematical work. The name of Descartes became known throughout Europe; Princes sought him; and it was only the outbreak of the civil war in England which prevented him from accepting a liberal appointment from Charles I. Instead, he went to Sweden at the invitation of Queen Christina, arriving at Stockholm in 1649, where it was hoped that he would found an Academy of Sciences. Such a replica of the Platonic School in Athens already existed in Paris. But his health gave way under the severity of the climate, and shortly after his arrival he died.

The work of Descartes changed the face of mathematics: it gave geometry a universality hitherto unattained; and it consolidated a position which made the differential calculus the inevitable discovery of Newton and Leibniz. For Descartes founded *analytical geometry,* and by so doing provided mathematicians with occupation lasting over two hundred years.

Descartes was led to his analytical geometry by systematically fitting algebraic symbols to the still fashionable rhetorical geometry. Examples of this procedure have already been given elsewhere. Those examples were stated in algebraic formulae in order to convey the sense of the propositions more readily to the reader. Strictly speaking, they were an anachronism before the time of Descartes. His next step concerned the famous Apollonian problem (p. 53) [*locus*] *ad tres et quattuor lineas,* disclosed by Pappus. It will be recalled that a point moves so that the product of its oblique distances from certain given lines is proportional to that of its distances from certain others. Descartes took a step that from one point of view was simplicity itself—he enlisted the fact

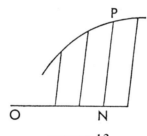

FIGURE 13

that plane geometry is *two*-dimensional. So he expressed everything in the figure in terms of two variable lengths, x and y, together with fixed quantities. This at once gave an algebraic statement for the results of Pappus: it put them into a form now typified by $f(x, y) = 0$, an equation where x and y alone are variable. The fundamental importance of this result lies in the further consequence that such an equation can be looked on as the definition of y in terms of x. It defined y as a function of x: it did geometrically very much what Napier's definition of a logarithm did dynamically. It also gave a new significance to the method of Archimedes for discussing the area of a curve, using an abscissa ON and an ordinate NP: in the notation of Descartes ON became x and NP, y. But, besides this, it linked the wealth of Apollonian geometry with what Archimedes had found; by forging this link Descartes rendered his most valuable service to mathematics.

Although Descartes deserves full credit for this, because he took considerable pains to indicate its significance, he was not alone in the discovery. Among others to reach the same conclusion was FERMAT—another of the great French mathematicians, a man of deeper mathematical imagination than Descartes. But Fermat had a way of hiding his discoveries.

Before indicating some of the principal consequences of this new method in geometry, there are other aspects of the notation which should be mentioned. The letter x has become world-famous: and it was the methodical Descartes who first set the fashion of denoting variables by x, y, z and constants by a, b, c. He also introduced indices to denote continued products of the same factor, a step which completed the improvements in notation originating with Diophantus. The fruitful suggestion of negative and fractional indices followed soon afterwards: it was due to WALLIS, one of our first great English mathematicians. A profound step in classification also was taken when Descartes distinguished between two classes of curves, *geometrical* and *mechanical,* or, as LEIBNIZ preferred to call them, *algebraic* and *transcendental.* By the latter is meant a curve, such as the spiral of Archimedes, whose Cartesian equation has no finite degree.

Apollonius had solved the problem of finding the shortest dis-

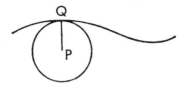

FIGURE 14

tance from a given point to a given ellipse, or other conic. Following this lead Descartes addressed himself to the same problem in general: he devised a method of determining the shortest line PQ from a given point P to a given curve. Such a line meets the curve at right angles in the point Q, and is often called the *normal* at Q to the curve. Descartes took a circle with centre P, and arranged that the radius should be just large enough for the circle to reach the curve. The point where it reached the curve gave him Q, the required foot of the normal. His way of getting the proper radius was interesting; it depended on solving a certain equation, two of whose roots were equal. It is hardly appropriate to go into further

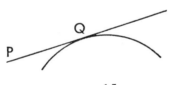

FIGURE 15

details here; but the reader who has some familiarity with analytical geometry, and has found the tangent to a circle or conic by the method of equal roots, has really employed the same general principle. Had Descartes been so inclined he could also have used his method for finding a tangent to a curve, i.e. a line PQ touching a given curve at a point Q (Figure 15). This is one of the first problems of the differential calculus; and one of the earliest solutions was found by Fermat and not by Descartes.

Fermat had discovered how to draw the tangent at certain points of a curve, namely at points Q which were, so to speak, at a crest or in the trough of a wave of the curve. They were points at a maximum or minimum distance from a certain standard base line called the axis of x. By so doing, Fermat

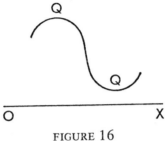

FIGURE 16

had followed up a fertile hint, which Kepler had let fall, concerning the behaviour of a variable quantity near its maximum or minimum values.

An interesting curve, still called the Cartesian oval, was discovered by Descartes, and has led to far-reaching research in geometry and analysis. It was found in an endeavour to improve the shape of a lens, so as to condense a pencil of light to an accurate focus. Although a lens of this shape would successfully focus a wide-angled pencil of light, if it issued from a certain particular position, the lens would be otherwise useless. But it has a physical, besides a mathematical, interest: for the principle underlying its construction is identical with that which Hero of Alexandria first noticed in the case of plane mirrors. It is the principle of Least Action, which was ultimately exhibited in a general form by Hamilton.

All this mathematical work was but part of a comprehensive philosophical programme culminating in a theory of vortices, by which Descartes sought to account for the planetary motions. Just as Kepler had thought of comets as live fishes darting through a celestial sea, Descartes imagined the planets as objects swirling in vast eddies. It remained for Newton not only to point out that this theory was incompatible with Kepler's planetary laws, but to propose a truer solution.

In philosophy Descartes made a serious attempt to build up a system in the only way which would appeal to a mathematician—by first framing his axioms and postulates. In doing this he was the true symbol of an age, filled with self-confidence after the triumphs of Copernicus, Napier and Kepler. We cannot but admire the intellectual force of a man who undertook to revise philosophy and achieved so much. Nevertheless he lacked certain gifts that might be thought essential to success in the venture. He was cold, prudent and selfish, and offered a great contrast to his younger contemporary, the mathematician and philosopher, Blaise Pascal.

The analytical geometry of Descartes is a kind of machine: and 'the clatter of the co-ordinate mill,' as Study has remarked, may be too insistent. The phenomenal success of this machine in the hands of Newton, Euler and Lagrange almost completely diverted thought from pure geometry. The great geometrical work in France, contemporary with that of Descartes, actually sank into oblivion for about two centuries, until it came into prominence once more, a hundred years ago. Two of the early French geometers were PASCAL and DESARGUES, and their work was the natural continuation of what Kepler had begun in projective geometry. Desargues, who was an engineer and architect residing at Lyons, gave to the ancient geometry of Apollonius its proper geometrical setting. He showed, for example, with grand economy, how to cut conics of different shapes from a single cone, and that a right circular cone. He won the admiration of Chasles, the great French geometer of the nineteenth century, who speaks of Desargues as an artist, but goes on to say that his work

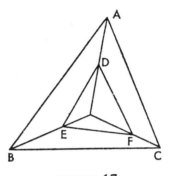

FIGURE 17

bears the stamp of universality uncommon in that of an artist. Desargues had the distinction of finding out one of the most important theorems of geometry, which takes its place, with a theorem of Pappus already quoted, as a fundamental element in the subject. It runs as follows: if two triangles ABC and DEF are such that AD, BE, CF meet in a point, then BC, EF; CA, FD; AB, DE, taken in pairs, meet in three points which are in line (Figure 17). The theorem is remarkable because it is easier to prove if the triangles are not in the same plane. As a rule, solid geometry is more difficult to handle than plane geometry—but not invariably. The perspective outline drawing of a cube on a sheet of paper is a more complicated figure than the actual outline of the solid cube. Desargues began the method of disentangling plane figures by raising them out of the flat into three dimensions. This is a choice method that has only lately borne its finest fruit in the *many-dimensional* geometry of Segre and the Italian school.

The work of Desargues is intimately linked with that of

Pascal. Even in the grand century which produced Descartes, Fermat and Desargues, the fourth great French mathematician, BLAISE PASCAL, stands out for the brilliancy of his genius and for his astonishing gifts. He was born at Clermont-Ferrand in Auvergne on 19th June, 1623; and was educated with the greatest care by his father, who was a lawyer and president of the Court of Aids. As it was thought unwise to begin mathematics too early, the boy was put to the study of languages. But his mathematical curiosity was aroused, when he was twelve years old, on being told in reply to a question as to the nature of geometry, that it consisted in constructing exact figures and in studying the relations between the parts. Pascal was doubtless stimulated by the injunction against reading it, for he gave up his playtime to the new study, and before long had actually deduced several leading properties of the triangle. He found out for himself the fact that the angles of a triangle are together equal to two right angles. When his father knew of it, he was so overcome with wonder that he wept for joy, repented, and gave him a copy of Euclid. This, eagerly read and soon mastered, was followed by the conics of Apollonius, and within four years Pascal had written and published an original essay on conic sections, which astounded Descartes. Everything turned on a miracle of a theorem that Pascal called 'L'hexagramme mystique,' commonly acknowledged to be the greatest theorem of mediaeval geometry. It states that, if a hexagon is inscribed in a conic, the three points of intersection of pairs of opposite sides always lie on a straight line: and from this proposition he is said to have deduced hundreds of corollaries, the whole being infused with the method of projection. The theorem has had a remarkably rich history, after the two hundred year eclipse, culminating in the enchantments of Segre when he presents it as a cubic locus in space of four dimensions, transfigured yet in its simplest and most inevitable form!

During these years Pascal was fortunate in enjoying the society in Paris of Roberval, Mersenne and other mathematicians of renown, whose regular weekly meetings finally grew into the French Academy. Such a stimulating atmosphere bore fruit after the family removed to Rouen, where at the age of eighteen Pascal amused himself by making his first calculating machine, and six years later he published his *Nouvelles Expériences sur le vide,* containing important experimental results which verified the work of Torricelli upon the barometer. Pascal was, in fact, as capable and original in the practical and experimental sciences as in pure geometry. At Rouen his father was greatly influenced by the Jansenists, a newly formed religious sect who denied certain tenets of Catholic doctrine, and in this atmosphere occurred his son's first conversion. A second conversion took place seven years later, arising from a narrow escape in a carriage accident. Henceforth Pascal led a life of self-denial and charity, rarely equalled and still more rarely surpassed. When one of his friends was condemned for heresy, Pascal undertook a vigorous defence in *A Letter written to a Provincial,* full of scathing irony against the Jesuits. Then the idea came to him to write an Apologia of the Christian Faith, but in 1658 his health, always feeble, gave way; and after some years of suffering borne with noble patience he died at the age of thirty-nine. The notes in which he jotted down his thoughts in preparation for this great project, have been treasured up and published in his *Pensées,* a literary classic.

In Pascal the simplest faith graced the holder of the highest intellectual gifts: and for him mathematics was something to be taken up or laid aside at the will of God. So when in the years of his retirement, as he lay awake suffering, certain mathematical thoughts came to him and the pain disappeared, he took this as a divine token to proceed. The problem which occurred to him concerned a curve called the cycloid, and in eight days he found out its chief properties by a brilliant geometrical argument. This curve may be described by the rotation of a wheel: if the axle is fixed, like that of a flywheel in a machine, a point on the rim describes a circle; but if the wheel rolls along a line, a point on the rim describes a cycloid. Galileo, Descartes and others were interested in the cycloid, but Pascal surpassed them all. To do so he made use of a new tool, the *method of indivisibles* recently invented by the Italian CAVALIERI. Though Pascal threw out a challenge, no one could compete with him: and his work may be regarded as the second chapter in the integral calculus, to which Archimedes had contributed the first.

An account of Pascal, the mathematician, would be incomplete without reference to his algebra, which, in the present-day sense of the word, he practically founded. It arose out of a game of chance that had formed a topic of discussion between Pascal and Fermat. From the debate the notion of mathematical *probability* emerged; this in turn Pascal looked upon as a problem in arrangements or combinations of given things and in counting those arrangements. With characteristic insight he lit upon the proper mechanism for handling the subject. It was the *Arithmetic Triangle*, a device already used by Napier for another purpose, and dating from still earlier times.

1	1	1	1	1	1
1	2	3	4	5	
1	3	6	10		
1	4	10			
1	5				
1					

Certain numbers are written down in a triangular table, as shown by the diagram. The table can at any stage be enlarged by affixing further numbers, one each at the right-hand extremities of the rows, with a single

1 added at the bottom of the first column to start a new row. For example, underneath the 5 of the second row, and alongside the 10 of the third row a new number can be placed. This number is 15, the sum of the 5 and the 10. According to this rule of simple addition each new number is entered in the table. The diagram exhibits a 1 in the top left-hand corner followed by five parallel diagonals, the fifth and last being (1, 5, 10, 10, 5, 1). A sixth, which has not been filled in, would consist of 1, 6, 15, 20, 15, 6, 1, according to the addition rule. Instead of locating an entry, 10 for example, as standing in the fourth row and third column, it is more important to locate it by the *fifth diagonal* and *third* column. Pascal discovered that this gave the number of combinations of *five* things taken *two* at a time; and he found a formula for the general case, when the number stood in the mth diagonal and the $(n + 1)$th column. He stated this correctly to be $(n + 1)(n + 2)(n + 3) \ldots (m)/1.2.3. \ldots (m - n)$. He also utilized the diagonals for working out the binomial expansion of $(a + b)^m$. For example,

$$(a + b)^5 = a^5 + 5a^4b + 10a^3b^2 + 10a^2b^3 + 5ab^4 + b^5.$$

Numbers and quantities are not always so important for their size or bulk as for their patterns and arrangements. What Pascal did was to bring this notion of pattern, common enough in geometry, to bear upon number itself—a highly significant step in the history of mathematics. By so doing he created higher algebra and prepared the way for Bernoulli, Euler and Cayley. 'Let no one say that I have said nothing new,' writes Pascal in his *Pensées;* 'the arrangement of the subject is new. When we play tennis, we both play with the same ball, but one of us places it better.'

FERMAT, who shared with Pascal the beginning of this algebra, is most famous for his theory of numbers. In the margin of a copy of Diophantus he made a habit of scribbling notes of ideas which came into his mind as he read. These notes are unique in their interest and profundity: he seemed to grasp properties of whole numbers by intuition rather than reason. The most celebrated note, which is often called *Fermat's Last Theorem*, has baffled the wit of all his analytical successors: for no one has yet been able to say whether Fermat was right or wrong. The theorem asserts that it is impossible to find whole numbers x, y, z which satisfy the equation

$$x^n + y^n = z^n$$

when n is an integer greater than 2. He adds: 'I have found for this a truly wonderful proof, but the margin is too small to hold it.' The problem has led to a wealth of new methods and new ideas about number; valuable prizes have been offered for a solution; but to-day its quiet challenge still remains unanswered.

Great things were also going on in Italy and England during this early seventeenth century. CAVALIERI of Bologna will always rank as a remarkable geometer who went far in advancing the integral calculus by his *Method of Indivisibles,* following up Kepler's wine-cask geometry. One of his theorems is a gem: upon concentric circles equally spaced apart he drew a spiral of Archimedes whose starting-point was the centre. Then in order to discover its area he re-drew the figure with all the circles straightened out into parallel lines the same distances apart as before. As a result the spiral became a parabola: and 'Unless I am mistaken,' he adds, 'this is a new and very beautiful way of describing a parabola' This is an early example of a transcendental mathematical transformation that not only preserves the area of a sector of the original curve but also the length of its arc.

Another very fine piece of work was done in 1695 by PIETRO MENGOLI, who gave an entirely new setting to the celebrated logarithm, by showing that it was intimately linked with a harmonical progression. His definition and treatment was on true Eudoxian lines and rigorous enough to satisfy the strictest arithmetical disciple of Weierstrass.

It is natural that, in these years succeeding Napier's death, a great deal of attention was bestowed upon the logarithm. Besides the practical business of constructing tables there was the still more interesting theory of logarithms to consider. The stimulus of analytical geometry encouraged several mathematicians to treat the logarithm by the method of co-ordinates. This led to a beautiful result that connected the area between a hyperbola and its asymptote with the logarithm. It was found in 1647 by GREGOIRE DE SAINT VINCENT, of Flanders: but several others turned their attention to the matter, reaching the same general conclusions more or less independently; notably Mercator, Mersennes, Brouncker, Wallis, James Gregory, Newton and Leibniz. (This Mercator was not the maker of geographical maps: he was a mathematician who had lived in the previous century.)

It is not difficult to suggest how this result was attained. A start was made with the geometrical progression whose sum is $1/(1-x)$; namely,

$$\frac{1}{1-x} = 1 + x + x^2 + x^3 + x^4 + \ldots,$$

and a curve was determined whose co-ordinate equation is $y = 1/(1-x)$. This curve is a hyperbola. Next, its area was determined, by following much the same course that Archimedes had taken for the case of the parabola. There was no difficulty in finding a requisite formula, thanks to Napier's original definition of the logarithm. It led to the result

$$\log (1 - x) = - x - \frac{x^2}{2} - \frac{x^3}{3} - \frac{x^4}{4} - \cdots,$$

which is called the logarithmic series. As may be seen, it is a union of the geometrical and harmonical progression.

Among the names which have just been given we find one Scot, one Irishman, and two Englishmen. For at last England produced mathematicians of the first rank, and in Gregory Scotland possessed a worthy successor to Napier. It is interesting to give, as typical specimens from the work of these our fellow-countrymen, the following formulae, which may be compared and contrasted with the logarithmic series:

$$\frac{4}{\pi} = \cfrac{1}{1 + \cfrac{1^2}{2 + \cfrac{3^2}{2 + \cfrac{5^2}{2 + \cdots}}}},$$

$$\frac{\pi}{4} = \frac{2 \times 4 \times 4 \times 6 \times 6 \times 8 \times \cdots}{3 \times 3 \times 5 \times 5 \times 7 \times 7 \times \cdots},$$

$$\frac{\pi}{4} = 1 - \tfrac{1}{3} + \tfrac{1}{5} - \tfrac{1}{7} + \cdots$$

The first is due to LORD BROUNCKER, an Irish peer; the second to WALLIS, who was educated in Cambridge and later became Savilian Professor of Mathematics in Oxford. The third was given by LEIBNIZ, but is really a special case of a formula discovered by JAMES GREGORY. Two of these formulae have been slightly altered from their original statements. The reader is not asked to prove, but merely to accept the results! After all, as they stand, they are readily grasped. The row of dots, with which each concludes, signifies that the formula can be carried farther; in fact, they each have something in common with the ladder-arithmetic of Athens (p. 29). They have this in common also with Vieta's formula for π (p. 67); but they improve on it, not only for their greater simplicity, but because each converges, as Plato would have it, by 'the great and small'—each step slightly overshooting the mark. This is not always done when such sequences are used, as in the more ordinary formula

$$\frac{\pi}{4} = \text{¼ of } 3 \cdot 1415926 \ldots = \cdot 785398 \ldots,$$

which approximates from one side only, like the putts of a timid golfer who *never* gives the ball a chance, or like the race of Achilles and the tortoise. Such series need careful handling, as Zeno had broadly hinted; and Gregory (by framing the notions of convergency and divergency) was the first to provide this.

In the last of these four formulae for $\frac{\pi}{4}$, the digits occur at random, and for this reason the statement is of little interest except to the practical mathematician. It is far otherwise with the other three: the *arrangement* of their parts has the inevitability of the highest works of art. It would be a pleasure to hear Pythagoras commenting upon them.

The Gregory family has long been associated with the county of Aberdeen. It had not been distinguished intellectually until John Gregory of Drumoak married Janet Anderson, herself a mathematician and a relative of the Professor of Mathematics in Paris. Many of their descendants have been eminent either as mathematicians or physicians. Chief among them all was their son James, who learnt mathematics from his mother. Unhappily, like Pascal, he died in his prime; but he lived long enough to exhibit his powers. After spending several years in Italy he occupied the Chair of Mathematics in St. Andrews for six years, followed by one year in Edinburgh. Shortly before his death he became blind.

Gregory was a great mathematical analyst, and many of his incidental results are striking. From the study of the logarithm he discovered the binomial theorem, generally and rightly attributed to Newton, who had probably found it out a few years earlier without publishing the result. It was but another case of independent discovery, as were also their invention of the reflecting telescope, and their attainments in the differential and integral calculus. The work of Gregory opened out a broad region of higher trigonometry, algebra and analysis. It is important not merely in detail theorems but for its general aim, which was to prove that no finite algebraic formula could be found to express the functions that arise in trigonometry and logarithms. In other words, he held that circle-squarers were pursuing a vainer phantom than those who endeavour with rule and compass to trisect an angle. His project was lofty, even if it inevitably failed: it was a brilliant failure in an attempt to disentangle parts of pure mathematics which were only satisfactorily resolved during the nineteenth century.

Some of his greatest work remained in manuscript until the Gregory tercentenary (1938) gave an opportunity to publish it. This included an important general theorem which was later discovered by Brook Taylor (1715). Paper was scarce in 1670 when Gregory used the blank spaces of old letters to record his work. This was the year when BARROW produced his masterpiece, the *Lectiones Geometricae,* in which the foundations of the differential and the integral calculus were truly but geometrically laid.

If it is asked what is the peculiar national contribution made by our country to mathematics, the reply is: the mathematics of interpolation—the mathematical art of reading between the lines. As an illustration let us consider the arithmetical triangle of Pascal, supposing it to be a fragment of an Admiralty chart. The numbers indicate the depth in fathoms at various points on the surface of the sea. Such a chart with these particular readings obviously indicates a submarine valley trending downwards south-east. What the chart does not show is the actual depth at positions intermediate between the readings. Mathematical interpolation is concerned with discovering a formula for the most probable depth consistent with these measured soundings. Certain isolated points are given: what is happening between? Napier, Briggs, Wallis, Gregory and Newton, each in his way gave an answer.

> *From gap to gap*
> *One hangs up a huge curtain so,*
> *Grandly, nor seeks to have it go*
> *Foldless ιand ʃflatɑ along the wall.*

Indeed, some faith was needed to believe that there *was* a curtain, and some imagination to see its pattern. For Napier it was the pattern of the logarithm; Wallis wrought a continuous chain out of the isolated exponents x^1, x^2, x^3, . . . , by filling in fractional indices. Newton found out the pattern which fills in the triangle of Pascal; and from this he discovered the binomial theorem in its general form. Briggs suggested and Gregory found an interpolation formula of very wide application, while Newton supplemented it with several other alternatives which have usually been attributed to Stirling, Bessel and Gauss.

CHAPTER VII

Isaac Newton

In the country near Grantham during a great storm, which occurred about the time of Oliver Cromwell's death, a boy might have been seen amusing himself in a curious fashion. Turning his back to the wind he took a jump, which of course was a long jump. Then he turned his face to the wind and again took a jump, which was not nearly so long as his first. These distances he carefully measured, for this was his way of ascertaining the force of the wind. The boy was ISAAC NEWTON, and he was one day to measure the force, if force it be, that carries a planet in its orbit.

From school at Grantham his friends took him to tend sheep and go regularly to the Grantham market. But as he *would* read mathematics instead of minding his business, it was at last agreed that he should go back to school, and from school to college. At school he lodged with Mr. Clark, apothecary, and in his lodgings spent much time, hammering and knocking. In the room were picture-frames and pictures of his own making, portraits, and drawings of birds and beasts and ships. Somewhere in the house might be seen a clock that was worked by water, and a mill which had a mouse as its miller. The boy made a carriage which could be propelled by the passenger, and a sundial that stood in the yard. To the little ladies of the house he was a very good friend, making tables and chairs for their dolls. His schoolfellows looked up to him as a skilful mechanic. As for his studies, when he first came to school he was somewhat lazy, but a fight that he had one day woke him up, and thereafter he made good progress. This quiet boy had great powers which were yet to be brought out.

In his twentieth year he went to Cambridge, where for more than thirty years he lived at Trinity College. He entered the college as a sizar, that is to say, being too poor to live in the style of other undergraduates he received help from the college. His tutor invited him to join a class reading Kepler's *Optics.* So Newton procured a copy of the book, and soon surprised the tutor by mastering it. Then followed a book on astrology; but this contained something which puzzled him. It was a diagram of the heavens. He found that, in order to understand the diagram, he must first understand geometry. So he bought Euclid's *Elements,* but was disappointed to find it too simple. He called it a 'trifling book' and threw it aside (an act of which he lived to repent). But turning to the work of Descartes he found his match, and by fighting patiently and steadily he won the battle.

After taking his degree Isaac Newton still went on learning all the mathematics and natural philosophy that Cambridge could teach him, and finding out new things for himself, until the Lucasian Professor of Mathematics in the University had become so convinced of the genius of this young man that, incredible as it may seem, he gave up to him his professorship. Isaac Barrow, the master of Newton's college, who thus resigned, was at no time a man to prefer self-interest before honour. He was possessed of great personal courage, and is reputed to have fought with a savage dog in an early morning's walk, and to have defended a ship from pirates. He was a mathematician of no mean powers; and as a divine he gained a lasting reputation.

Newton made three famous discoveries: one was in light, one was in mathematics, and one in astronomy. We are not to suppose that these flashed upon him all at once. They were prepared for by long pondering. 'I keep,' said he, 'the subject of my inquiry constantly before me, and wait till the first dawning opens gradually, by little and little, into a full and clear light.' Early in his career he discovered that white light was composed of coloured lights, by breaking up a sunbeam and making the separate beams paint a rainbow ribbon of colours upon a screen. This discovery was occasioned by the imperfection of the lenses in telescopes as they were then made. Newton chose to cure the defect by inventing a reflecting telescope with a mirror to take the place of the principal lens, because he found that mirrors do not suffer from this awkwardness of lenses. It is one of his distinctions, shared with Archimedes and a few other intellectual giants, that his own handiwork was so excellent. In the chapel of his college there is a statue, holding a prism:

> —*Newton, with his prism and silent face;*
> *The marble index of a mind for ever*
> *Voyaging through strange seas of thought, alone.*

In mathematics his most famous discovery was the differential and integral calculus—which he called the method of *Fluxions:* and in astronomy it was the conception and elaboration of universal gravitation. It would be a mistake to suppose that he dealt with these subjects one by one: rather they were linked together, and reinforced each other. Already at the age of twenty-three, when for parts of the years 1665 and 1666 the college was shut down owing to the plague, Newton had thought out, in his quiet country home, the principles of gravitation and, for the better handling of the intense mathematical difficulties which the principles involved, he had worked at the fluxional calculus. In the space of three years after first reading geometry, he had so completely mastered the range of mathematics from Archimedes to Barrow, that he had fitted their wonderful infinitesimal geometry into a systematic discipline. Newton gave to analysis the same universality that Descartes had already given to geometry.

Newton may be said to have fused the points of view adopted by Napier and Descartes into a single whole. Napier thought of points M and N racing along parallel tracks OX and OY, N moving steadily and M at a variable speed. The co-ordinates of Descartes provide a chart of the race in the following way: the lines OX and OY can be placed, no longer parallel, but at right angles to each other, and a curve can be plotted, traced by a point P which is simultaneously abreast of the points N and M. In this way two figures can be drawn, one the Naperian and the other the Cartesian. The figures are symbols of two lines of thought—the kinematical and the geometrical. Newton may never actually have drawn such figures side by side, but he certainly had the two trains of thought. 'I fell by degrees on the method of fluxions,' he remarks: and by *fluxions* he simply meant what we call the simultaneous speeds of the points N and M. Then by seeking to compare the speed of M with that of N, he devised the method which the geometrical figure suggests. 'Fluxions' was his name for what we call the differential and integral calculus, but he kept the discovery to himself.

In after years Leibniz announced that *he* had found this new mathematical method. Then a quarrel arose between the followers of Newton and the followers of Leibniz, and unhappily it grew into a quarrel between the great men themselves. It is enough to say that the time was ripe for such a discovery: and both Newton and the German philosopher were sufficiently gifted to effect it. Newton was the first to do so, and only brought the trouble unwittingly on his head by refraining from publishing his results. It is also probable that

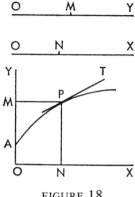

Leibniz was influenced more by Pascal and Barrow than by Newton: and in turn we owe to Leibniz the record of parts of Pascal's work which would otherwise have been lost.

About this time the Royal Society of London was founded by King Charles II. It corresponded with the Academy of Paris, and provided a rendezvous for the leading mathematicians and natural philosophers in the country. Two of the Fellows of this Society were Gregory and Newton, who had become friends through their common interest in the reflecting telescope. Besides maintaining a correspondence, they may actually have met. They were certainly brought into contact with other leading mathematicians and astronomers. Among those who have not already been mentioned in the last chapter were Wren, Hooke, and Halley.

Christopher Wren is now so famous as the builder of St. Paul's Cathedral that we never hear of his scientific fame, though a man of science he was. Hooke, who was in appearance a puny little man, was a hard student, often working till long after midnight, but caring too excessively for his own reputation. When Newton found out anything, Hooke would commonly remark, 'That is just what I found out before.' But he was a great inventor, whose eager speculations stirred people up to think about the questions which Newton was to solve. Halley was an astronomer—a very active man, always travelling about the world to make some addition to his science. Every one has heard of Halley's comet, and to Halley is due the credit of bringing Newton before the world as the discoverer of gravitation.

One day these three friends were talking earnestly together: the subject of their conversation was the whirlpool theory of Descartes, which they felt to be hardly a satisfactory explanation of planetary motion. It did not seem to give a proper explanation of the focal position of the Sun within the elliptic orbit. Instead of imagining the planets to be propelled by a whirling current, they preferred to think that each planet was forcibly attracted by the Sun. 'Supposing,' said they, 'the Sun pulls a planet with such and such a force, how ought the planet to go? We want to see clearly that the planet will go in an ellipse. If we can see that, we shall be pretty sure that the Sun *does* pull the planet in the way we supposed.' 'I can answer that,' said Hooke: upon which Wren offered him forty shillings on condition of his producing the answer within a certain time. However, nothing more was heard of Hooke's solution. So at last after several months Halley went to Cambridge, to consult Newton; and, without mentioning the discussion which had taken place in London, he put the question: If a planet were pulled by the Sun with a force which varies inversely as the square of the distance between them, in what sort of a curve ought the planet to go? Newton, to Halley's astonishment and delight answered, 'An ellipse.' 'How do you know that?' 'Why, I have calculated it.' 'Where's the calculation?' Oh, it was somewhere among his papers; he would look for it and send it to Halley. It appeared that Newton had worked all this out long before; and only now in this casual way was the matter made known to the world. Then Halley did a wise thing: he persuaded his retiring friend to develop the entire problem, explaining the whole complicated system of planetary motion. This Newton did; it was a tremendous task, taking two or three years; at the end of which appeared the famous book called *The Mathematical Principles of Natural Philosophy,* or more shortly the *Principia,* one of the supreme achievements of the human mind.

It is impossible to exaggerate the importance of the book, which at once attracted the keenest attention not only in England but throughout Europe. It was a masterpiece alike of mathematics and of natural philosophy. Perhaps the strangest part of the work was not so much the conception that the Sun pulls the planet, but that the planet pulls the Sun—and pulls equally hard! And that the whole Universe is full of falling bodies: and everything pulls everything else—literally everything, down to the minutest speck of dust. When Newton's friends had discussed the effect of the solar attraction upon a planet, they had correctly surmised the requisite force: it was determined by what is called the *law of the inverse square.* Newton had already adopted this law of force in his early conjectures, during the long vacation of 1666, over twenty years before the publication of the *Principia* (1687). That early occasion is also the date to which the well-known apple story may be referred. It is said that the sight of a falling apple set in motion the train of thought, leading Newton to his discovery of universal gravitation. But after working out the mathematical consequences of his theory and finding them to disagree with the observed facts he had tossed his pages aside. Only after many years he became aware of later and more careful calculations of the observations. This time, to his delight, they fitted his mathematical theory, and so Newton was ready with his answer, when Halley paid him the memorable visit.

In the *Principia* Newton demonstrated that, if his rule of gravitation is universally granted, it becomes the key to all celestial motions. Newton could not *prove* that it was the right key, for not all the celestial motions were known at the time, but very nearly all that have since been discovered help to prove that he was right. Even so, there was enough already known to give Newton plenty of trouble. The moon, for instance, that refuses to go round the Earth in an exact ellipse, but has all sorts of fanciful little excursions of her own—the moon was very trying to Isaac Newton.

Newton's great book was written in Latin, and, in order to make it intelligible to current habits of mind, it was couched in the style of Greek geometry. Newton had of course worked the mathematics out by fluxions, but he preferred to launch the main gravitational discovery alone, without further perplexing his readers by the use of a novel method. Outside his Cambridge lecture-room little was known of his other mathematical performances until a much later date. His *Arithmetica Universalis* was published in 1707, and two more important works, on algebra and geometry, appeared about the same time. Newton left his mark on every branch of mathematics which he touched; indeed, there are few parts of the subject which escaped his attention. Allusion has already been made to his work in interpolation and algebra. The power of his methods may be judged from one celebrated theorem which he gave without proof for determining the positions of the roots of an equation. A hundred and fifty years elapsed before Sylvester discovered how to prove his theorem.

The publication of the *Principia* forced Newton to abandon his sheltered life. In 1689 he became a Member of Parliament, and a few years later was appointed Master of the Mint. In

1705 he was knighted by Queen Anne. He died in 1727 at an advanced age, and was buried in Westminster Abbey. Voltaire has recorded his pride at having lived for a time 'in a land where a Professor of Mathematics, only because he was great in his vocation, was buried like a king who had done good to his subjects.' The world at large is often more generous in showing appreciation and gratitude than are mathematicians themselves, who feel, but are slow to exhibit their feelings. It is therefore the more noteworthy that, two hundred years later, in 1927, the English mathematical world made a pilgrimage to Grantham, to signal their respect for the genius of Newton. This alone is enough to indicate that the immense reputation which he always enjoyed was fully deserved.

It is proper to associate, with Newton, the great Dutch natural philosopher HUYGENS (1629–1693), who was in close touch with scientists of England, and did much to stimulate their wonderful advances. His own work in physics is so grand that his mathematics are apt to be overlooked. He contributed many elegant results in the infinitesimal calculus, particularly in its bearings upon mechanical phenomena, the oscillations of a pendulum, the shape of a hanging string, and the like. But he is best known for his undulatory theory of light.

As a mathematical concept this has proved to be a landmark in the history, and it is particularly interesting because it has thrown Newton's universal gravitation into intense relief. To Newton light seemed to be so many tiny particles streaming in luminous lines: to Huygens, on the contrary, light was propagated by waves. The sequel has shown that, of these rival theories, the latter is the more valuable. Not only has it provided a better key to optical puzzles, but it has also answered many purposes in the theory of electricity and magnetism. One by one, the natural phenomena were absorbed

105

in this all-enveloping wave-theory, and gravitation alone remained untouched—a single physical exception. This unwave-like behaviour of gravitation, this action at a distance, sorely perplexed Newton himself, long before these further instances of natural behavior had made wave motion the correct deportment. As the mystery of gravitation deepened, it became more and more the conscious aim of scientists to explain the contrast: and the matter has only lately been settled by Einstein, who solves the problem by drastically embedding gravitation in the very texture of space and time.

But it would be wrong to suppose that this left the field clear for the wave-theory. Quietly and unobtrusively other interpretations have been congregating, and reasons have once more been urged in favour of Newton's corpuscular theory of light. At the present time there is no clear cut decision one way or the other: the work of both Newton and Huygens appears to be fulfilled in the Quantum Theory and the Wave Mechanics.

CHAPTER VIII

The Bernoullis and Euler

The story of mathematics during the eighteenth century is centred upon Euler, and the scene of action is chiefly laid in Switzerland and Russia. About the time when Napier was experiencing the turmoil of the Reformation, violent persecution of Protestants took place in Antwerp. One of the many refugees, whom Belgium could ill afford to lose, was a certain Jacques Bernoulli, who fled to Frankfort. In 1622 his grandson settled at Basel, and there, on the frontiers of Switzerland, the BERNOULLI family were destined to bring fame to the country of their adoption. As evidence of the power of heredity, or of early home influence, their mathematical record is unparalleled. No less than nine members of the family attained eminence in mathematics or physics, four of whom received signal honours from the Paris Academy of Sciences. Of these nine the two greatest were the brothers JACOB and JOHN, great-grandsons of the fugitive from Antwerp. Jacob was fifth child in the large family, and John, thirteen years his junior, was tenth. Each in turn became Professor of Mathematics at Basel.

The elder brother settled to his distinguished career, as a mathematical analyst, only after considerable experiment and travel. At one time his father had forbidden him to study either mathematics or astronomy, hoping that he would devote himself to theology. But an inborn talent urged his son to spend his life in perfecting what Pascal and Newton had begun. Among his many discoveries, and perhaps the finest of them all, is the equiangular spiral. It is a curve to be found in the tracery of the spider's web, in the shells upon the shore and in the convolutions of the far-away nebulae. Mathematically it is related in geometry to the circle and in analysis to the logarithm. A circle threads its way over the radii by crossing them always at right angles; this spiral also crosses its radii at a constant angle—but the angle is not a right angle. Wonderful are the phœnix-like properties of the curve: let all the mathematical equivalents of burning it and tearing it in pieces be performed—it will but reappear unscathed! To Bernoulli in his old age the curve seemed to be no unworthy symbol of his life and faith; and in accordance with his wishes the spiral was engraved upon his tombstone, and with it the words *Eadem mutata resurgo*.

His younger brother JOHN (1667–1748) followed in his footsteps, continually adding fresh material to the store of analysis which now included differential equations. His work exhibits a bolder use of negative and imaginary numbers, thereby realizing 'the great emolument' which Napier himself had hoped to bestow on mathematics by 'this ghost of a quantity,' had not his own attention been absorbed by logarithms. His sons Daniel and Nicolas Bernoulli were also very able mathematicians, and it was under their influence at college that Euler discovered his vocation.

LEONARD EULER (1707–1783) was the son of a clergyman who lived in the neighbourhood of Basel. His natural aptitude for mathematics was soon apparent from the eagerness and facility with which he mastered the elements under the tuition of his father. At an early age he was sent to the University of Basel, where he attracted the attention of John Bernoulli.

Inspired by such a teacher he rapidly matured, and at the age of seventeen, when he received the degree of Master of Arts, he provoked high applause for a probationary discourse, the subject of which was a Comparison between the Cartesian and Newtonian Systems.

His father earnestly wished him to enter the ministry and directed his son to study theology. But unlike the father of Bernoulli, he abandoned his views when he saw that his son's talents lay in another direction. Leonard was allowed to resume his favourite pursuits and, at the age of nineteen, he transmitted two dissertations to the Paris Academy, one upon the masting of ships, and the other on the philosophy of sound. These essays mark the beginning of his splendid career.

About this time, in consequence of the keen disappointment at failing to attain a vacant professorship in Basel, he resolved to leave his native country. So in 1727, the year when Newton died, Euler set off for St. Petersburg to join his friends, the younger Bernoullis, who had preceded him thither a few years earlier. On the way to Russia, he learnt that Nicolas Bernoulli had fallen a victim to the stern northern climate; and the very day upon which he set foot on Russian soil the Empress Catherine I died—an event which at first threatened the dissolution of the Academy, of which she had laid the foundation. Euler, in dismay, was ready to give up all hope of an intellectual career and to join the Russian navy. But, happily for mathematics, when a change took place in the aspect of public affairs in 1730, Euler obtained the Chair of Natural Philosophy. In 1733 he succeeded his friend Daniel Bernoulli, who wished to retire; and the same year he married Mademoiselle Gsell, a Swiss lady, the daughter of a painter who had been brought to Russia by Peter the Great.

Two years later, Euler gave a signal example of his powers, when in three days he effected the solution of a problem urgently needed by members of the Academy, though deemed insoluble in less than several months' toil. But the strain of the work told upon him, and he lost the sight of an eye. In spite of this calamity he prospered in his studies and discoveries, each step seeming only to invigorate his future exertions. At about the age of thirty he was honoured by the Paris Academy when he received recognition, as also did Daniel Bernoulli and our own countryman Colin Maclaurin, for dissertations upon the flux and reflux of the sea. The work of Maclaurin contained a celebrated theorem upon the equilibrium of elliptical spheroids; that of Euler brought the hope considerably nearer of solving outstanding problems on the motions of the heavenly bodies.

In the summer of 1741 King Frederick the Great invited Euler to reside in Berlin. This invitation was accepted, and until 1766 Euler lived in Germany. On first arriving he received a royal letter written from the camp at Reichenbach, and he was soon after presented to the queen-mother, who always took a great interest in conversing with illustrious men. Though she tried to put Euler at his ease, she never succeeded in drawing him into any conversation but that of monosyllables. One day when she asked the reason for this, Euler replied, 'Madam, it is because I have just come from a country where every person who speaks is hanged.' It was during his residence in Berlin that Euler wrote a remarkable set of letters, or lessons, on natural philosophy, for the Princess of Anhalt Dessau, who was eager for instruction from so great a teacher. These letters are a model of perspicuous and interesting teaching, and it is noteworthy that Euler should have found time for such detailed elementary work, amid all his other literary interests.

For eleven years his widowed mother lived in Berlin also, receiving assiduous attention from her son, and enjoying the pleasure of seeing him universally esteemed and admired. Euler became intimate in Berlin with M. de Maupertuis, President of the Academy, a Frenchman from Brittany who strongly favoured Newtonian philosophy in preference to Cartesian. His influence was important, as it was exerted at a time when Continental opinion was still reluctant to accept the views of Newton. Maupertuis much impressed Euler with his favourite principle of least action, which Euler used with great effect in his mechanical problems.

It speaks highly for the esteem in which Euler was held that, when in 1760 a Russian army invaded Germany and pillaged a farm belonging to Euler, and the act became known to the general, the loss was immediately made good, and a gift of four thousand florins was added by the Empress Elizabeth when she learnt of the circumstance. In 1766 Euler returned to Petersburg, to spend the remainder of his days, but shortly after his arrival he lost the sight of his other eye. For some time he had been forced to use a slate, upon which in large characters he would make his calculations. Now, however, his pupils and children copied his work, writing the memoirs exactly as Euler dictated them. Magnificent work it was too, astonishing at once for its labour and its originality. He developed an amazing facility for figures, and that rare gift of mentally carrying out far-reaching calculations. It is recorded that on one occasion when two of his pupils, working the sum of a series to seventeen terms, disagreed in their results by one unit at the fiftieth significant figure, an appeal was made to Euler. He went over the calculation in his own mind, and his decision was found to be correct.

In 1771, when a great fire broke out in the town and reached Euler's house, a fellow-countryman from Basel, Peter Grimm, dashed into the flames, discovered the blind man and carried him off on his shoulders into safety. Although books and furniture were all lost, his precious writings were saved. For twelve years more Euler continued his excessive labours, until the day of his death, in the seventy-sixth year of his age.

Like Newton and many others, Euler was a man of parts, who had studied anatomy, chemistry and botany. As is reported of Leibniz, he could repeat the *Aeneid* from beginning to end, and could even remember the first and last lines in every page of the edition which he had been accustomed to use. The power seems to have been the result of his most wonderful concentration, that great constituent of inventive power, to which Newton himself has borne witness, when the senses are locked up in intense meditation, and no external idea can intrude.

Sweetness of disposition, moderation and simplicity of manner were his characteristics. His home was his joy, and he was fond of children. In spite of his afflicton he was lively and cheerful, possessed of abundant energy; as his pupil M. Fuss has testified, 'his piety was rational and sincere; his devotion was fervent.'

In an untechnical account it is impossible to do justice to the mathematics of Euler: but while Newton is a national hero, surely Euler is a hero for mathematicians. Newton was the Archimedes and Euler was the Pythagoras. Great was the work of Euler in the problems of physics—but only because their mathematical pattern caught and retained his attention. His delight was to speculate in the realms of pure intellect, and here he reigns a prince of analysts. Not even geometry, not even the study of lines and figures, diverted him: his ultimate and constant aim was the perfection of the calculus and analysis. His ideas ran so naturally in this train, that even in Virgil's poetry he found images which suggested philosophic inquiry, leading on to new mathematical adventures. Adventures they were, which his more wary followers sometimes hailed with delight and occasionally condemned. The full splendour of the early Greek beginnings and the later works of Napier, Newton and Leibniz, was now displayed. Let one small formula be quoted as an epitome of what Euler achieved:

$$e^{i\pi} + 1 = 0.$$

Was it not Felix Klein who remarked that all analysis was centred here? Every symbol has its history—the principal whole numbers 0 and 1; the chief mathematical relations $+$ and $=$; π the discovery of Hippocrates; i the sign for the 'impossible' square root of minus one; and e the base of Napierian logarithms.

CHAPTER IX

Maclaurin and Lagrange

Among the contemporaries of Euler there were many excellent mathematicians in England and France, such as Cotes, Taylor, Demoivre, D'Alembert, Clairaut, Stirling, Maclaurin, and, somewhat later, Ivory, Wilson and Waring. This by no means exhaustive list contains the names of several friends of Newton—notably Cotes, Maclaurin and Demoivre. They were Newton's disciples, and each was partly responsible for making the work of the Master generally accessible. Cotes and Maclaurin were highly gifted geometers: the others of their time were interested in analysis. It was therefore a loss not only to British but to European mathematics that Cotes and Maclaurin should both have died young.

COLIN MACLAURIN (1698–1746), a Highlander from the county of Argyle, was educated at the University of Glasgow. Such was his outstanding ability that, at the age of nineteen, he was elected Professor of Mathematics in Aberdeen. Eight years later, when he acted as deputy Professor in Edinburgh, Newton wrote privately offering to pay part of the salary, as there was difficulty in raising the proper sum. Maclaurin took an active part in opposing the march of the Young Pretender in 1745 at the head of a great Highland army, which overran the country and finally seized Edinburgh. Maclaurin escaped, but the hardships of trench warfare and the subsequent flight to York proved fatal, and in 1746 he died.

Stirred by the brilliant work of Cotes, which luckily came

into his hands, Maclaurin wrote a wonderful account of higher geometry. He dealt with the part which is called the *organic description of plane curves,* a subject belonging to Euclid, Pappus, Pascal and Newton. It is the mathematics of rods and bars, constrained by pivots and guiding rails—the abstract replica of valve gears and link motions familiar to the engineer —and it fascinates the geometer who 'likes to see the wheels go round.' Maclaurin carried on what Pascal had begun with, the celebrated mystic hexagram (which at that date still lay hid), and in so doing he reached a result of great generality. It provided a basis for the advances in pure geometry that were made a century later by Chasles, Salmon and Clifford. In this kind of geometry the Cartesian method of co-ordinates fails to keep pace with the purely geometrical. In it men breathe a rarer air, akin to that in the theory of numbers.

The very success of Maclaurin partakes of the tragic. For there are huge tracts of mathematics where co-ordinates provide the natural medium—where, for any but a supreme master, analysis succeeds and pure geometry leaves one helpless. When Maclaurin wrote his essay on the equilibrium of spinning planets, which gained him the honours of the Paris Academy, he set out on a course wherein few could follow: for the problem was rendered in the purest geometry. When, in addition to this, Maclaurin produced a great geometrical work on fluxions, the scale was so heavily loaded that it diverted England from Continental habits of thought. During the remainder of the century British mathematics were relatively undistinguished, and there was no proper revival until the differential calculus began to be taught in Cambridge, according to the methods of Leibniz—a change which took place about a hundred years ago. This delay was the unhappy legacy of the Newton-Leibniz controversy, which need never have arisen.

The circumstances that prompted Maclaurin to adopt a geometrical style in his book on fluxions, extended beyond his partiality for geometry. Many philosophical influences were at work, and there were logical difficulties to face, which seemed to be insurmountable except by recourse to geometry. The difficulties were focused on the word *infinitesimal*—which Eudoxus had so carefully excluded from the vocabulary of Greek mathematics (the mere fact that it is a Latin, and not a Greek, word is not without its significance; so many of our ordinary mathematical terms have a Greek derivation). By an infinitesimal is meant something, distinguishable from zero, yet which is exceedingly small—so minute indeed that *no* multiple of it can be made into a finite size. It evades the axiom of Archimedes. Practically all analysts, from Kepler onwards, believed in the efficacy of infinitesimals, until Weierstrass taught otherwise. The differential calculus of Leibniz was founded on this belief, and its tremendous success, in the hands of the Bernoullis, Euler and Lagrange, obscured the issue. Men were disinclined to reject a doctrine which worked so brilliantly, and they turned a deaf ear to the philosophers, ancient and modern. In our own country a lively attack on infinitesimals was headed by the Irish philosopher and theologian, Bishop Berkeley. His criticism of the calculus was not lost upon Maclaurin, who was also well versed in Greek mathematics and the careful work of Eudoxus. So Maclaurin made up his mind to put fluxions upon a sound basis and for this reason threw the work into a geometrical frame. It was his tribute to Newton, the master 'whose caution,' said Maclaurin, 'was almost as distinguishing a part of his character as his invention.'

One of the chief admirers of Maclaurin was Lagrange, the great French analyst, whose own work offered a complete contrast to that of the geometer. Maclaurin had dealt in lines and figures—those characters, as Galileo has finely said, in which the great book of the Universe is written. Lagrange, on the contrary, pictured the Universe as an equally rhythmical theme of numbers and equations; and was proud to say, of his masterpiece, the *Mécanique Analytique,* that it contained not a single geometrical diagram. Nevertheless he appreciated the true geometer, declaring that the work of Maclaurin surpassed that of Archimedes himself, while as for Newton, he was 'the greatest genius the world has ever seen—and the most fortunate, for only once can it be given a man to discover the system of the Universe!'

JOSEPH-LOUIS LAGRANGE (1736–1813) came of an illustrious Parisian family which had long connnection with Sardinia, and some trace of noble Italian ancestry. He spent his early years in Turin, his active middle life in Berlin, and his closing years in Paris, where he attained his greatest fame. Foolish speculation on the part of his father threw Lagrange, at an early age, upon his own resources, but this change of fortunes proved to be no great calamity, 'for otherwise,' he says, 'I might never have discovered my vocation.' At school his boyish interests were Homer and Virgil, and it was not until a memoir of Halley came his way, that the mathematical spark was kindled. Like Newton, but at a still earlier age, he reached to the heart of the matter in an incredibly short space of time. At the age of sixteen he was made Professor of Mathematics in the Royal School of Artillery at Turin, where the diffident lad, possessed of no tricks of oratory and very few words, held the attention of men far older than himself. His winning personality elicited their friendship and enthusiasm. Very soon he was conducting a youthful band of scientists who became the earliest members of the Turin Academy. With a pen in his hand Lagrange was transfigured; and from the first, his writings were elegance itself. He would set to mathematics all the little themes on physical inquiries which his friends brought him, much as Schubert would set to music any stray rhyme that took his fancy.

At the age of nineteen he won fame by solving the so-called isoperimetrical problem, that had puzzled the mathematical world for half a century. He communicated his proof in a letter to Euler, who was immensely interested in the solution, particularly as it agreed with a result that he himself had found. With admirable tact and kindness Euler replied to Lagrange, deliberately withholding his own work, that all the credit might fall on his young friend. Lagrange had indeed not only solved a problem, he had also invented a new method, a new *Calculus of Variations,* which was to be the central subject of his life-work. This calculus belongs to the story of Least Action, which began with the reflecting mirrors of Hero (p. 50) and continued when Descartes pondered over his curiously shaped oval lenses. Lagrange was able to show that the somewhat varied Newtonian postulates of matter and motion fitted in with a broad principle of economy in nature. The principle has led to the still more fruitful results of Hamilton and Maxwell, and it continues today in the work of Einstein and in the latest phases of Wave Mechanics.

Lagrange was ready to appreciate the fine work of others, but he was equally able to detect a weakness. In an early memoir on the mathematics of sound, he pointed out faults even in the work of his revered Newton. Other mathematicians ungrudgingly acknowledged him first as their peer, and later as the greatest living mathematician. After several years of the utmost intellectual effort he succeeded Euler in Berlin. From time to time he was seriously ill from overwork. In Germany King Frederick, who had always admired him, soon grew to like his unassuming manner, and would lecture him for his intemperance in study which threatened to unhinge his mind. The remonstrances must have had some effect, because Lagrange changed his habits and made a programme every night of what was to be read the next day, never exceeding the ration. For twenty years he continued to reside in Prussia, producing work of high distinction that culminated in his *Mécanique Analytique*. This he decided to publish in France, whither it was safely conveyed by one of his friends.

The publication of this masterpiece aroused great interest, which was considerably augmented in 1787 by the arrival in Paris of the celebrated author himself, who had left Germany after the death of King Frederick, as he no longer found a sympathetic atmosphere in the Prussian Court. Mathematicians thronged to meet him and to show him every honour, but they were dismayed to find him distracted, melancholy, and indifferent to his surroundings. Worse still—his taste for mathematics had gone! The years of activity had told; and Lagrange was mathematically worn out. Not once for two whole years did he open his *Mécanique Analytique:* instead, he directed his thoughts elsewhere, to metaphysics, history, religion, philology, medicine, botany, and chemistry. As Serret has said, 'That thoughtful head could only change the objects of its meditations.' Whatever subject he chose to handle, his friends were impressed with the originality of his remarks. His saying that chemistry was 'easy as algebra' vastly astonished them. In those days the first principles of atomic chemistry were keenly canvassed: but it seemed odd to draw a comparison between such palpable things as chemicals, that can be handled and seen, and such abstractions as algebraic symbols.

In this philosophical and unmathematical state of mind Lagrange continued for two years, when suddenly the country was plunged into the Revolution. Many avoided the ordeal by flight abroad, but Lagrange refused to leave. He remained in Paris, wondering as he saw his friends done to death if *his* turn was coming, and surprised at his good fortune in surviving. France has reason to be glad that he was not cut down as was his friend Lavoisier, the great chemist; for in later years mathematical skill once again returned to him, and he produced many gems of algebra and analysis.

One mathematical effect of the Revolution was the adoption of the metric system, in which the subdivision of money, weights and measures is strictly based on the number ten. When someone objected to this number, naturally preferring twelve, because it has more factors, Lagrange unexpectedly remarked what a pity it was that the number eleven had not been chosen as base, because it was prime. The M.C.C. appears to be one of the few official bodies who have followed this hint, by thinking systematically in terms of such a unit!

For music he had a liking. He said it isolated him and helped him to think, as it interrupted general conversation. 'For three bars I listen to it; thereafter I distinguish nothing, but give myself up to my thoughts. In this way I have solved many a difficult problem.' He was twice married: first when he lived in Berlin, where he lost his wife after a long illness, in which he nursed her devotedly. Then again in Paris he married Mlle. Lemonnier, daughter of a celebrated astronomer. Happy in his home life, simple and almost austere in his tastes, he spent his quiet fruitful years, till he died in 1813 at the age of seventy-six.

Lagrange is one of the great mathematicians of all time, not only for the abundance and originality of his work but for the beauty and propriety of his writings. They possess the grandeur and ease of the ancient geometers, and Hamilton has described the *Mécanique Analytique* as 'a scientific poem.' He was equally at home rivalling Fermat in the theory of numbers and Newton in analytical mechanics. Much of the contemporary and later work of Laplace, Legendre, Monge,

Fourier and Cauchy, was the outcome of his inspiration. Lagrange sketched the broad design; it was left to others to fill in the finished picture. One must turn to the historians of mathematics to learn how fully and completely this was done. The breadth of the canvas attracted men of widely different interests. Nothing could afford a greater contrast to the mind of Lagrange than that of Laplace, the other great contributor to natural philosophy, whose most notable work was the *Mécanique Céleste*. To Laplace mathematics were the accidents and natural phenomena the substance—a point of view exactly opposite to that of Lagrange. To Laplace mathematics were tools, and they were handled with extraordinary skill, but any makeshift of a proof would do, provided that the problem was solved. It remained for the nineteenth century to show the faultiness of this naïve attitude. The instinct of the Greeks was yet to be justified.

CHAPTER X

Gauss and Hamilton: The Nineteenth Century

The nineteenth century, which links the work of Lagrange with that of our own day, is perhaps the most brilliant era in the long history of mathematics. The subject assumed a grandeur in which all that was great in Greek mathematics was fully recovered; geometry once again came into its own, analysis further broadened its scope, and the outlets for its applications were ever enlarging. The century was marked in three noteworthy ways: there was deeper insight into the familiar properties of number; there was positive discovery of new processes of calculation, which, in the quaint words of Sylvester, ushered in 'the reign of Algebra the Second'; and there was also a philosophy of mathematics. During these years England once again rivalled mathematical France, and Germany and Italy rose to positions of scientific importance; while pre-eminent over all was the genius of one man, a mathematician worthy of a place of honour in the supreme rank with Archimedes and Newton.

CARL FRIEDRICH GAUSS was born in 1777 at Brunswick, and died in 1855, aged seventy-eight. He was the son of a bricklayer, and it was the wish of his father that he should be a bricklayer too. But at a very early age it was clear that the boy had unusual talents. Unlike Newton and Lagrange he showed the precocity of Pascal and Mozart. It is said that Mozart wrote a minuet at the age of four, while Gauss pointed out to his father an error in an account when he was three. At school his cleverness attracted attention, and eventually

122

became known to the Duke of Brunswick himself, who took an interest in the lad. In spite of parental protest the Duke sent him for a few years to the Collegium Carolinum and in 1795 to Göttingen. Still undecided whether to pursue mathematics or philology, Gauss now came under the influence of Kaestner—'that first of geometers among poets, and first of poets among geometers,' as the pupil was proud to remark. In the course of his college career Gauss became known for his marvellous intuiton in higher arithmetic. 'Mathematics, the Queen of the Sciences, and Arithmetic, the Queen of Mathematics,' he would say: and mathematics became the main study of his life.

The next nine years were spent at Brunswick, varied by occasional travels, in the course of which he first met his friend Pfaff, who alone in Germany was a mathematician approximating to his calibre. After declining the offer of a Chair at the Academy in St. Petersburg, Gauss was appointed in 1807 to be first director of the new observatory at Göttingen, and there he lived a studious and simple life, happy in his surroundings, and blessed with good health, until shortly before his death. Once, in 1828, he visited Berlin, and once, in 1854, he made a pilgrimage to be present at the opening of the railway from Hanover to Göttingen. He saw his first railway engine in 1836, but except for these quiet adventures, it is said that until the last year of his life he never slept under any other roof than that of his own observatory!

His simple and direct character made a profound impression upon his pupils, who, seated round a table and not allowed to take notes, would listen with delight to the animated address of the master. Vivid accounts have been handed down of the chief figure in the group as he stood before his pupils 'with clear bright eyes, the right eyebrow raised higher than the left (for was he not an astronomer?), with a forehead high and wide, overhung with grey locks, and a countenance whose variations were expressive of the great mind within.'

Like Euler, Lagrange and Laplace, Gauss wrote voluminously, but with a difference. Euler never condensed his work; he revelled in the richness of his ideas. Lagrange had the easy style of a poet; that of Laplace was jerky and difficult to read. Gauss governed his writings with austerity, cutting away all but the essential results, after taking endless trouble to fill in the details. His pages stimulate but they demand great patience of the reader.

Gauss made an early reputation by his work in the theory of numbers. This was but one of his many mathematical activities, and, apart from all that followed, it would have placed him in the front rank. Like Fermat, he manifested that baffling genius which leaps—one knows not how—to the true conclusion, leaving the long-drawn-out deductive proof for others to formulate. A typical example is provided by the *Prime Number Theorem* which has taken a century to prove. Prime numbers were studied by Euclid, and continue to be an eternal source of interest to mathematicians. They are the numbers, such as 2, 3, 5, 7, 11, that cannot be broken up into factors. They are infinitely numerous, as Euclid himself was aware, and they occur, scattered through the orderly scale of numbers, with an irregularity that at once teases and captivates the mathematician. The question is naturally suggested: *How often, or how rarely, do prime numbers occur on the average?* Or, put in another way, What is the chance that a specified number is prime? In some form or other this problem was known to Gauss; and here is his innocent-looking answer:

"Primzahlen unter a ($= \infty$)

$$\frac{a}{la}.$$"

It means that when *a* is a very large number, the result of dividing *a* by its logarithm gives a good approximation to the total number of primes less than *a:* and the larger *a* is, the more precise is the result. Whether Gauss proved his statement is not known: the quotation is taken from the back page of a copy of Schulze's Table of Logarithms which came into his possession when he was fourteen. Probably he recorded his note a few years later. Even if we recall the history of the logarithm and its diverse relations with so many remote parts of mathematics, this latest example is not a little strange. The contents of a book of logarithms, with its thickly crowded tables of decimal fractions, appears to be foreign indeed to the delicate distribution of primes among the whole numbers.

An actual proof of this theorem was given only as recently as thirty years ago by Hadamard and de la Vallée Poussin. It is an example of a new and very abstract part of the subject, now called the analytical theory of numbers. This is one of the striking developments in the present century, that has been notably advanced in Germany by Landau and in our country by Hardy and Littlewood.

Ever since the time of Gauss, mathematics has increased so extensively that no individual could hope to master the whole. Gauss was the last complete mathematician, and of him it can truly be said that he adorned every branch in the science. The beginnings of nearly all his discoveries are to be found in the youthful notes that he jotted down in a diary, unmethodically kept for several years, which has happily been preserved. The diary reveals pioneer facts in higher trigonometry, a subject generally known as Elliptic Functions: it also contains certain aspects of non-Euclidean geometry.

There is no doubt that Gauss was led to take an interest in geometry through Kaestner, his master, who himself had written on the fundamentals of the subject. Another contemporary influence was that of Legendre, whose book, the *Élé-ments de géometrie*, had appeared in 1794. These authors were interested in a problem that had often been discussed, notably by Wallis in England, and Saccheri, an Italian monk of the early eighteenth century. It concerned the parallel postulate of Euclid—that curious rugged feature in the smooth logic of the ancients, the removal of which seemed so very desirable. Gauss was perhaps the first to offer a satisfactory explanation of the anomaly: and his diary shows how early in his career this occurred. But, like Newton he was a cautious man, particularly when he handled strange and disconcerting novelties. For some years he kept the matter to himself, until he found that others were thinking of the same things. Among his college friends was a Hungarian, W. Bolyai, with whom he still corresponded; and in 1804 Bolyai wrote him a letter bearing on this theory of parallels. The interest spread, and out of it grew a branch of geometry called hyperbolic geometry. This branch of the subject is now always associated with the names of Gauss and his two friends the Bolyais, father and son, and of Lobatchewski, a Russian, who wrote some twenty years later. It is another case of several independent discoveries on one theme all taking place in the same era.

Hyperbolic geometry was not merely a novelty; it was a revolution. In a very practical way it ran counter to Euclid, and in a still more practical way it ran counter to the current opinions of what Euclid was supposed to teach. Euclid declared, for instance, that the sum of the three angles of a

FIGURE 19

triangle is equal to two right angles. He also declared that the sum of two adjacent angles, made by cross lines, is equal to two right angles. Both of these properties were implied, as he showed, in his fundamental axioms and postulates. But according to Gauss and Bolyai, while the declaration about the cross lines is true, that about the triangle is not: in fact they fashioned a triangle for which the sum of the angles is *less* than two right angles. Then, by nice balance, a little later, RIEMANN and others did the same for a triangle in which the sum is *greater* than two right angles. They called theirs elliptic geometry: it is the geometry that sailors know so well who voyage in direct courses over curved oceans of the globe. Less, equal and greater: three contradictory statements. These give rise to three bodies of geometrical doctrine: elliptic, parabolic and hyperbolic, the parabolic being that of Euclid. Here was the making of a first-class battle, not between opposing scientific camps holding relatively vague conflicting hypotheses, but in the very stronghold of logical argument—the realm which every one had taken for granted as settled and secure. The battle was fought, and it came to an end. As a three-cornered contest all sides lost in this sense, that any partisan for one of the views, saying *this* is true and both the others are false, would find himself pursuing a wild-goose chase. Instead of holding sovereign sway, each of these three is found to subserve a more fundamental whole. The programme of last century was designed to unravel the essential from the unessential, to isolate and underline right reasons for each geometrical fact, stripped of all misleading lumber. If, for example, we speak of the straight line AB, in the direction AB, we are not speaking redundantly, but of two different things, straightness and direction. Admittedly this is very puzzling, but it is none the less a fact.

An illustration of these abstract ideas is very literally ready to hand. Every one knows that it is easy to fix a small piece of plaster to the back of the hand, but quite difficult to fix it over a knuckle or between two knuckles. In these cases the plaster has to be crumpled or stretched, to make it adhere. This has a mathematical explanation. The back of the hand presents a surface agreeable to Euclidean geometry, but the knuckles and the gaps do not. The knuckles illustrate Riemann, and the gaps, Gauss. In the gap a triangular piece of plaster would have its angles crumpled and therefore *less* than two right angles; it would need to be elastic, and expand, in order to fit over a knuckle. There is nothing very difficult in apprehending these ideas, because the surface of the hand is two-dimensional. When, however, the same notions are applied to space itself, as inevitably they were, the mind refuses readily to accommodate itself to the effort. A necessary prelude was the study of the sticking-plaster type of geometry, and this was done by Gauss. He developed the theory of surfaces, with special attention to their curvature and the conditions for one surface to fit another. It is said that he laid aside several questions which he treated analytically, and hoped to apply to them geometrical methods in some future state of existence, when his conceptions of space should have become amplified and extended.

Riemann, one of his many celebrated pupils, partly fulfilled this aspiration of Gauss. He certainly improved analysis out of all knowledge by his ingenious geometrical interpretation for the theory of functions. Also, in a few pages of epoch-making dissertation, he not only contemplated geometry for space of any dimensions—a surmise that he shared with Cayley—but showed that the earlier three types of geometry were particular instances of a still more general geometry. If geometry is likened to the surface of a sea, then these three correspond to the surface in a calm; that of Riemann corresponds to the surface in a calm or storm. His thesis was the necessary prelude to that of Einstein.

Another great pupil of Gauss was an Irish Rugby schoolboy of the days of Arnold, HENRY J. S. SMITH by name. He became Savilian Professor of Mathematics at Oxford, and handed on the tradition of Gauss in the theory of numbers. Even among mathematicians, the highly original work of Smith on the borderland of arithmetic and algebra is hardly as well known as it ought to be: for he originated certain important developments which brought fame to others, notably Weierstrass, Frobenius and Kronecker.

Smith owed much to his talented mother, who was early left a widow, schooling her children with abundant leisure in comparative isolation, to grow up like the young Brontës in a world of their own. He became an excellent linguist, in doubt at first whether to follow mathematics or classics, and it is said of him that no British mathematician ever came nearer to the spirit of the ancient Greek philosophers. As a boy he had something of the wisdom of a man, and to the day of his death he retained the simplicity and high spirits of a boy.

Ireland has produced many great mathematicians: and another was SALMON, who did so much to reconcile the geometry of Pascal and Descartes, and whose books have been an education in themselves. Later, when he became a distinguished theologian, he showed the same power and lucidity in his theological writings that had marked his mathematics. But the greatest figure of all was WILLIAM ROWAN HAMILTON, who made two splendid discoveries, an early one in optics, on the Principle of Least Action, and later the Quaternions in algebra. He was born in 1805, and educated at Trinity College, Dublin, where at the age of twenty-one he became Professor of Astronomy, continuing to hold the office until 1865, the year of his death. He was a poet, and a friend of Wordsworth and Coleridge, and between these three passed a highly interesting correspondence, dealing with philosophy, science, and literature.

As a child, Hamilton astonished every one by his early powers. At three he could read English; at four he was thoroughly interested in geography and had begun to read Latin, Greek and Hebrew; before he was ten he had slaked his thirst for Oriental languages by forming an intimate acquaintance with Sanscrit, and grounding himself in Persian, Arabic, Chaldee, Syriac and sundry Indian dialects. Italian and French were imbibed as a matter of course, and he was ready to give vent to his feelings in extemporized Latin. Taking to this monumental programme with ease and diligence, he was for all that a vigorous child, as ready to romp and run and swim as any other small boy.

In his seventeenth year he began to think for himself upon the subject of optics, and worked out his great principle of the *Characteristic Function* which, four years later, he presented to the Irish Academy in a thesis entitled an *Account of a Theory of Systems of Rays*. This youthful production was a work of capital importance in natural philosophy, as may be gathered from the sequel. Along with certain work in electromagnetism by Clerk Maxwell, it shares the hard-won distinction of triumphantly surviving the latter-day revolution caused by the theory of Relativity.

Owing to its importance in the history of mathematics a quotation from Hamilton's thesis may not be inappropriate. After noticing how others—and particularly Malus, an officer who served under Napoleon—had invoked the principle of least action in studying rays of light, he says:

'A certain quantity which in one physical theory is the *action* and in another the *time*, expended by light in going from any first to any second point, is found to be less than if the light had gone in any other than its actual path. . . . The mathematical novelty of .my method consists in considering this quantity as a function . . . and *in reducing all researches respecting optical systems of rays to the study of this single function:* a reduction which presents mathematical optics under an entirely novel view, and one analogous (as it appears to me) to the aspect under which Descartes presented the application of algebra to geometry.'

So light navigates space, as sailors navigate an ocean, by seeking the direct path. It is therefore hardly surprising that the work of Gauss and Hamilton should have eventually merged in a broader mathematical harmony. It needed but one step more—to devise a means of applying these ideas to space of *more* than the three ordinary dimensions—for the theory of Relativity to come into being. This essential step was taken by Christoffel, who worked with Riemannian geometry, and to-day the grandly sounding *World Function* of Hilbert is none other than the Characteristic Function of the youthful Hamilton, rehabilitated in four dimensions. It was a stroke of genius when Einstein found in this exceedingly elaborate geometry the very medium needed to cope with actual physical phenomena.

Hamilton also had high views on algebra, which he designated 'the science of pure time,' and in making his discovery of *quaternions* brought into being an entirely new method of computation. Though his quaternions behaved very like numbers, they were not numbers; for they broke the commutative law. By this is meant the law under which it is asserted that $2 \times 3 = 3 \times 2$ or $a \times b = b \times a$ for ordinary numbers. As this law had gently been appropriated, at each stage, for all the new types of number, fractional, negative, irrational and indeed complex—the finishing touches of Gauss and Cauchy being, so to say, hardly dry—the mathematical world was lulled into slumber; little expecting something explosive to occur in *this* quarter. Nevertheless an explosion came—two explosions, in fact. One was fired by Hamilton and the other in Germany by GRASSMAN. For each discovered independently the need, in geometry or dynamics, of algebraic symbols whose behaviour was exemplary, judged by all accepted numerical standards—except the commutative law. For such symbols, the products ij and ji differed. If, says Hamilton, $ij = k$ then $ji = -k$.

Hamilton made this discovery in 1843 at the age of thirty-eight. It came like a flash, to relieve an intellectual need that had haunted him for fifteen years. Already Möbius had invented a sort of geometrical weighing machine which he called the *Barycentric Calculus,* whereby, not merely numbers, but points and forces could be *added* together. Out of this grew the notion of vectors, a name given to cover diverse physical phenomena such as forces and velocities. Hamilton called his vectors triplets, because forces act in three dimensions, and in course of time he was anxious to find a way for their multiplication. His home circle became interested in this puzzle. Every morning, on coming down to breakfast, one of his little boys used to ask, 'Well, Papa, can you multiply triplets?' Whereto he was obliged to answer with a sad shake of the head, 'No, I can only *add* and subtract them.' But one day, so he relates, he was walking with his wife beside the Royal canal on his way to a meeting in Dublin of the Academy. Although she talked with him now and then, yet an undercurrent of thought was going on in his mind, which at last gave a result. It came in a very tangible form, and it at once suggested to him many a long year of purposeful work upon an important theme. He could not resist the impulse to cut with a knife, on the stone of Brougham Bridge, as they passed it, the fundamental formulae

$$i^2 = j^2 = k^2 = ijk = -1,$$

indicative of the *Quaternions* which gave the solution of the problem.

This instinctive feeling that the discovery was important was well founded. Hamilton and Grassmann provided the first examples of a vast range in mathematics to which the same *algebra* has been appropriated. Arithmetic, algebra, analysis and geometry—these are the ingredients of Mathematics, each with its influence on the others, yet each with its own peculiar flavour. It was one of the characteristics of last century to emphasize the peculiarities, so that now we have a much clearer notion of their several significances. The subjects have been present from the beginnings of the science, and Eudoxus, Pythagoras, Archimedes, and Apollonius were forerunners of these several branches: Eudoxus with his interest in pure number, Pythagoras for his patterns and arrangements of things, Archimedes for his speculations about the infinite, and Apollonius for his projections of lines and curves.

CHAPTER XI

More Recent Developments

The discovery of quaternions by Sir William Hamilton was a signal for the revival of mathematics throughout the country. Henceforward through the nineteenth century not only Ireland, but England and Scotland were once again represented in the foremost rank. A very prominent group of English mathematicians included Boole, Cayley and Sylvester, all of whom made important contributions to the new algebra inaugurated by Hamilton. BOOLE, who came originally from Lincolnshire but spent many of his most active years in Ireland, made an important discovery, not unlike that of the quaternions. He found that it was possible to apply algebraic symbols to logic, a step that went far towards clarifying our fundamental ideas in both logic and mathematics. He was also the pioneer in the algebraic theory of invariants, for in 1841 he discovered the first specimen of such a function. Out of this grew the work of CAYLEY and SYLVESTER, two of the greatest British mathematicians of the century. Cayley brought mathematical glory to Cambridge, second only to that of Newton, and the fertility of his suggestions, in geometry and algebra, continues to influence the whole range that is now studied at home and abroad. To this versatility Cayley added a Gauss-like care and industry.

Cayley was a sage, but his friend, the fiery, enthusiastic Sylvester, was a poet. At one time they both resided in London, where they were studying for a legal profession, in days before a happy turn of fortune drew them into their truer avocations. Many an important addition to higher algebra was made, as they strolled round the Law Courts eagerly discussing invariants. Subsequently Cayley returned to his old University, becoming Sadlerian Professor of Mathematics in Cambridge. For a time Sylvester lived in America, fostering an algebraic tradition, the fruit of which is one of the features of twentieth-century mathematics. The ground had already been prepared by PIERCE of Harvard, who played the rôle, in America, of Hamilton and Grassmann in Europe.

It is noteworthy that in 1925 another example occurred of an abstract mathematical theory providing the mechanism for a new physical development. This took place when Heisenberg found, in the algebra of Hamilton as generalized by Cayley, the key to his new mechanics. To-day the subject is known as the Wave Mechanics, and it has received several different treatments. In one of these, which is due to Schrödinger, the Characteristic Function of Hamilton reappears, as a natural vehicle to explain the heart-beats of the atom.

Our ancestors in the Middle Ages received a shock when it was found that the surface of the earth, boundless as it appeared to them, was limited and could be circumnavigated. A similar shock was felt, in the narrower world of mathematics, when in 1868 it was found that a certain set of algebraic expressions, or invariants, that appeared to be endless, was finite. The credit of this startling discovery rests with GORDAN of Erlangen, a small university town of South Germany—a place already famous for the geometer, Von Staudt, and his able successors. This theorem of Gordan led HILBERT, a few years later, to formulate his *Basis Theorem,* of extraordinarily wide application, which can be regarded as giving a sort of algebraic blessing to the quaternions of Hamilton and Grassmann, as well as to large tracts of actual arithmetic. Gordan's

135

own proof had involved a long piece of mathematical induction, marvellously handled. That of Hilbert, on the contrary, was short and depended on such general principles that it drew from Gordan the comment, 'This is theology, not mathematics!'

The geometer VON STAUDT shares with Grassmann a belated fame, for their contemporaries quite failed to realize the profundity and originality of their work. Von Staudt belongs to a distinguished group of geometers, among whom was found once again the spirit of Pappus, Pascal and Desargues. His name is singled out for special mention because he alone in the group considered, and successfully dealt with, fundamental principles. In abandoning the geometry of Euclid and adopting one or other of the non-Euclidean systems, it is difficult to avoid a feeling of being cast hopelessly adrift on a sea of chance. But the work of Von Staudt, as much as that of any man, has rendered a sober discussion of non-Euclidean geometry possible; for he has revealed the solid foundations common to all these types of geometry. His 'unpretentious but imperishable little volume' has led us to apprehend what is fundamental and what is not. For example, the first concern in geometry is whether three points A, B, C are in line or not, quite apart from whether B lies between A and C, or how far apart they are: these are doubtless important, but *secondary,* considerations. This is good theory; but after all it is also good common sense. It is easier on a level plain to get two distant posts in line with the eye, than to discover which of them is nearer, or exactly how far off they are. The really remarkable thing about it all is that Pappus and Desargues had actually hit upon the fundamental theorems of geometry (pp. 53, 85) in spite of using proofs involving unnecessary assumptions. It is just as if they had drawn their figures in red ink and had thought that they would not be valid in any other colour. Nineteenth-century mathematics was largely concerned with getting rid of that red ink.

The developments of Analysis indicate the same general principles at work. Analysis is the branch of mathematics dealing with the infinite, the unbounded—either the immeasurably large or the immeasurably small. After Gauss, prodigious advances were made. Perhaps the most important of these was that of WEIERSTRASS and the Berlin school, which finally settled the Newton-Leibniz controversy by a return to the ancient methods of Eudoxus. In geometry, Pappus and Desargues had given wrong reasons for the right results, and the same thing frequently happened in the calculus. It may be recalled that Zeno had cured the Greeks of any such loose reasoning and that his criticisms had led to the epoch-making work of Eudoxus. But never since the Renaissance had this been properly assimilated, although Wallis, Newton and Maclaurin came nearest to doing so. The task of reconciliation was undertaken by Weierstrass, and also by DEDEKIND of Göttingen. They said, in effect, that analysis has to do with number —not with geometry; therefore let a strictly arithmetical account be given. In this they succeeded, both for the calculus and for the theory of irrational numbers; and two of their principal means, in attaining this end, were the definitions of irrationals due to Eudoxus, and those of limits given by Wallis and Newton.

Euler and his contemporaries had provided an armoury of analytical weapons: they were now rendered keen and refined. Many a famous old problem consequently fell before the cunning onslaughts of the analysts. One of the most spectacular results was that of LINDEMANN, who proved that the irrational number π satisfies no algebraic equation with integer coefficients. This settled the matter of squaring the circle, once and for all.

With such a piling up of analytical armaments it might be thought that all simplicity had forsaken mathematics. At the close of the century nothing could have seemed more desirable than the rise of a genius who could dispense with all these elaborations, and yet find something new to say. Very dramatically this took place in India, and the career of SRINIVASA RAMANUJAN has marked a new epoch. India has from time to time possessed mathematicians of great power: they may be traced through the ages back to the later Greek period. But judged by absolute standards of greatness, among all mathematicians of the East, the genius of Ramanujan appears to be supreme. He was born at Erode, a town not far from Madras, in 1887 and died in 1920. At school his extraordinary powers seem to have been recognized, but owing to his weakness in English he failed to matriculate at the University of Madras. He therefore proceeded to work out mathematics for himself, deriving what help he could from Carr's *Synopsis of Pure Mathematics*. After several years' work in the Harbour Board office at Madras he became known to someone sufficiently interested in the contents of his mystifying notebooks to put him in touch with the mathematical experts. Notes that were locally unintelligible received immediate recognition in Cambridge as the work of a self-taught genius. An invitation to visit Cambridge was accepted, and in due time Ramanujan became a Fellow of Trinity College and of the Royal Society, in token of his conspicuous merit. Unhappily the residence in England destroyed his health, and the year after his return to India he died.

Difficult as it is to form a judgment of almost contemporary work, there can be no question that here was an exceptional mathematician. In spite of all the disadvantages of his mathematical upbringing, with its scanty supply of material, he attained a command of certain branches in analysis and the theory of numbers that placed him in the very front rank, even before he was discovered by the West. Sylvester once took Huxley to task for thinking that 'mathematics is the study which knows nothing of observation, nothing of experiment, nothing of induction, nothing of causation.' Such a description completely misrepresents mathematics, which in its making unceasingly calls forth the highest efforts of imagination and invention. As for induction, there is no more wonderful instance of its use than that of Ramanujan. One of his problems happened to have been solved independently by Landau; yet, as Hardy has remarked, Ramanujan 'had none of Landau's weapons at his command; he had never seen a French or German book. It is sufficiently marvellous that he should have even dreamt of problems such as these, problems which have taken the finest mathematicians of Europe a hundred years to solve, and of which the solution is incomplete to the present day.' Nevertheless there were curious blind spots in his work where it went definitely wrong. Traversing many of the roads taken by Wallis, the Bernoullis and Euler, who occasionally blundered on to the wrong path in their enterprising adventures, Ramanujan repeated in his own short career the experiences of three centuries. Yet 'with his memory, his patience, and his power of calculation, he combined a power of generalization, a feeling for form, and a capacity for rapid modification of his hypotheses, that was often really startling, and made him, in his own peculiar field, without a rival in his day.'

Perhaps his greatest monument is a theorem that he discovered jointly with Hardy, dealing with the partitions of a number *n*. The theorem determines the number of ways in which *n* can be expressed as a sum of smaller whole numbers: and once more the simplicity of the enunciation completely disguises the profound difficulty of the quest. The theorem was a genuine example of collaboration, involving characteristic leaps in the dark, that border on the miraculous, followed by searching applications of Western mathematical analysis. As Littlewood has said: 'We owe the theorem to the happy collaboration of two men of quite unlike gifts, in which each contributed the best, most characteristic, and most fortunate work that was in him. Ramanujan's genius did have this one opportunity worthy of it.'

There were still other very general concepts that were brought into mathematics during the latter half of last century, particularly the theory of groups, which dates from the student days in Paris of two friends, SOPHUS LIE and FELIX KLEIN, of Scandinavia and Germany; and the theory of assemblages created by GEORG CANTOR, a Dane. Both of these ideas have had enormous influence on the trend of recent and contemporary thought; it is enough to say here that they exhibit in diverse ways the more mathematical side of that philosophical search into the principles of our subject which has marked the most recent stage of its history. For mathematics had now reached a state in which it was possible to do for the whole what Euclid tried to do for geometry, by disclosing the underlying axioms or primitive propositions, as Peano called them: and the most patient investigation has been made—notably in our own land by Whitehead and Russell—first of the subject-matter itself and next of the very ideas that govern the subject-matter. As all this was conceived on a sublimely universal scale, it is hardly remarkable that certain paradoxes have come to light. How to face these paradoxes is an urgent problem, and there are at the present time two or three different schools of thought employed upon this. The school associated with Brouwer of Holland has adopted a most drastic policy. They trace the presence of paradoxes to the use of indirect proofs, or more precisely to what is called in logic the law of the excluded middle. To this they object, very much as others have earlier objected to the Parallel Postulate of Euclid; indeed, it may be a symptom of the advent of a high synthesis in arithmetic and analysis, just as the earlier was in geometry. As nothing less than the whole edifice from Eudoxus to Cantor is at stake, little wonder that these views cause a stir in the mathematical world. 'Of what use,' said Kronecker to Lindemann, 'is your beautiful investigation regarding π? Why study such problems, since irrational numbers are non-existent?' So back we are once more at a logical scandal such as troubled the Greeks. The Greeks survived and conquered it, and so shall we. At any rate, it is all a sign of the eternal freshness of mathematics.

*   *   *   *   *

The story has now been told of a few among many whose admirable genius has composed the lofty themes which go to form our present-day heritage. Agelong has been the noble toil that has called forth a simplicity and steadfastness of purpose in all its greatest exponents. And if this little book perhaps may bring to some, whose acquaintance with mathematics is full of toil and drudgery, a knowledge of those great spirits who have found in it an inspiration and delight, the story has not been told in vain. There is a largeness about mathematics that transcends race and time: mathematics may humbly help in the market-place, but it also reaches to the stars. To one, mathematics is a game (but what a game!) and to another it is the handmaiden of theology. The greatest mathematics has the simplicity and inevitableness of supreme poetry and music, standing on the borderland of all that is wonderful in Science, and all that is beautiful in Art. Mathematics transfigures the fortuitous concourse of atoms into the tracery of the finger of God.

141